JOB MATTERS

SANITÄR- UND KLIMATECHNIK

von
Wolfram Lepka
Peter Oldham
Ingrid Preedy
Ken Thomson

unter Mitarbeit der Verlagsredaktion

Cornelsen

How to use the book

Job Matters Sanitär- und Klimatechnik wurde für die Ausbildung in Berufsschulen sowie die innerbetriebliche Aus- und Weiterbildung konzipiert und bietet Ihnen praxisorientiertes Fachenglisch basierend auf den Anforderungen des Gemeinsamen Europäischen Referenzrahmens (A2-B1).

So finden Sie sich im Buch zurecht:

Zu Beginn jeder Unit finden Sie eine Übersicht über die Themen und sprachlichen Fertigkeiten, die in der jeweiligen Unit behandelt werden.

IT'S YOUR TURN

Die ***It's your turn***-Aufgaben am Ende jeder Unit stellen einen zusammenfassenden Abschluss der Unit dar und richten sich an diejenigen, die gezielt an weiterführenden sprachlichen Fertigkeiten arbeiten möchten. Hier können zudem eigene Erfahrungen der Lernenden eingebracht werden.

▶▶▶ ICH KANN

	Seite/Übung
☐ Ich kann eine eingeübte, kurze und einfache Präsentation zu einem Thema vortragen, das mir vertraut ist. Ich kann dabei einfache Nachfragen beantworten, wenn ich um Wiederholung bitten kann. A2	25/ It's your turn

Nach jeder vierten Unit greifen ***Extra practice***-Seiten das Gelernte noch einmal auf und bieten so ein zusätzliches Übungsangebot zu den Themen der vorherigen Units. Außerdem zeigen ***Can-do-Statements***, welche Fertigkeiten die Lernenden in den Units erworben und gefestigt haben.

Für heterogene Klassen werden innerhalb der einzelnen Units differenzierte Aufgaben angeboten und durch Sterne gekennzeichnet (★ = niedriger Schwierigkeitsgrad, ★★ = mittlerer Schwierigkeitsgrad, ★★★ = hoher Schwierigkeitsgrad).

(Company name) … speaking. Can I help you? / How can I help you?
I can be with you in half an hour / an hour / … minutes / … Is that okay?
I can come over at … (time) / this morning/afternoon. Would that be convenient?
I have a problem with …
Would it be possible for you to come now? / It's a bit of an emergency. / When would you be able to call by?

Praxisbezogene Redewendungen und Phrasen finden Sie in den grünen Kästen.

Weitere wichtige Symbole:

2 = kennzeichnet Höraufgaben und gibt den entsprechenden Track an. Die Hörbeispiele sind separat als Download (ISBN 9783064513358) unter www.cornelsen.de erhältlich.

= kennzeichnet Tipps und wichtige Hinweise

Viel Spaß beim Lernen und gutes Gelingen wünschen Ihnen die Autoren und der Verlag!

Contents

1 Your career

This unit is about:
- basic skills
- a plumber's tasks

Competences:
- listening to a plumber talking about work
- talking about typical tasks
- creating a mind map of typical jobs in plumbing

1 Choosing a career in plumbing

Interview one of your classmates. Find out why he/she chose a career in plumbing.

Here are some ideas for questions:

Why did you choose a career in plumbing?

How did you find a job?

Is there a plumber in your family?

Do you enjoy your job?

What tasks do you have to do?

What do you think a good plumber must be good at?

Here are some ideas for answers:

I am good at / enjoy working with my hands. I like installing/repairing things.
My careers teacher told me about the job. I saw a job advert in the paper / on the Internet.
My uncle/father/mother/grandfather is a plumber.
I enjoy my job. / I don't enjoy my job. / I like my job a lot.
I have to clean the workshop / clean tools / install dishwashers, washing machines, etc. / deal with customers / install water systems / work with heating systems / …
A good plumber must be hard-working / good at maths / good at working with his/her hands / …

2

2 A popular trade

Brendan Mason owns a large plumbing and heating company in London. In a radio interview he is talking about how to become a plumber.

★ a) Listen to the recording and then fill in the missing words.

Interviewer: In today's programme, "A career as a tradesman", we are talking to master plumber Brendan Mason about how to become a plumber. Hello Brendan, welcome to our programme.

Brendan: Hello, thank you for inviting me.

Interviewer: Brendan, do you think becoming a plumber is a good ________________ [1] for young people?

Brendan: Yes, I believe that becoming a plumber is an ideal career opportunity for ________________ [2] young people who like to work with their hands.

Interviewer: What exactly does a plumber do?

Brendan: A plumber provides a wide ________________ [3] of services, such as installing and maintaining various types of piping systems and ________________ [4]. The exact responsibilities of a plumber depend on the particular project type. Plumbers may also design piping systems for new ________________ [5] projects.

Interviewer: What skills should a young person have if he or she wants to become a plumber?

Brendan: He or she must be able to understand basic ________________ [6], have a head for heights, be able to work in cramped conditions and be able to do ________________ [7] hard work day after day.

Interviewer: And how do you become a plumber?

Brendan: An ________________ [8] with a recognized firm will give you on-the-job training and teach you about the different types of pipes, the correct use of ________________ [9] and how to read plumbing plans. At the same time you will learn the theory at a vocational college on a three-year part-time course.

★★ b) Work with a partner. What does a plumber do? Make a list of useful vocabulary and present it in class.

3 A plumber's tasks

Match the typical plumbing tasks in the box with the pictures below.

checking pipes for a blockage caused by plants • installing a fire sprinkler system • installing air conditioning • installing a shower • installing a washing machine • installing pipes on a building site • repairing central heating • unblocking a toilet

1

____

2

____

3

____

4

____

5

____

6

____

7

____

8

____

IT'S YOUR TURN

Make a mind map of typical jobs that you have to do during your training. Which of these jobs do you think is the most interesting? Discuss with a partner and find reasons for your choice.

2 Customer service

This unit is about:
- a telephone conversation
- politeness

Competences:
- listening to a recording
- doing a role-play
- understanding good customer service

3 ## 1 Dealing with customers

Good customer service is important when dealing with customers.

★ a) **Jeff Norman calls Wilkie's Plumbing Services and talks to one of the plumbers there, Barry. Listen to the telephone call and answer the questions below.**

1 Why is Jeff Norman calling?

____

2 Why can't Barry answer his question?

____

3 What does Barry say he will do to find out the answer to the question?

____

4 What happened the last time Wilkie's Plumbing Services did a job for Mr Norman?

____

____

5 Why does Barry say he will call Mr Norman back?

____

____

★★ b) **Now listen again and tick (✔) the actual sentences that are used in the dialogue. The other sentences are also acceptable ways of saying the same things.**

How can I help you? ☐
What can I do for you? ☐
I'm sorry, but I don't know anything about that, so I can't answer your question. ☐
Oh, I'm sorry, I'm afraid I'm not working on that job, so I can't answer your question right away. ☐
Give me a few minutes and I'll ask a colleague to let you know. ☐
But if you give me a few minutes, I can ask a colleague who will be able to tell me. ☐
I'm sorry to hear that there was a problem. ☐
I really am sorry that you had a problem. ☐
I'm certain that our men won't be late tomorrow. ☐
I'm sure that our men will do their best to be on time tomorrow. ☐
Can I call you back in a few minutes, please? ☐
I'll give you a call back in a few minutes. ☐

★★★ c) **Work with a partner. Take on the two roles and act out similar dialogues. Here are some ideas for changes.**

Tom Jordan	Good morning	in two hours/days	wife	kitchen sink
Mabel Taylor	Good afternoon	at 10.30 a.m.	husband	shower
Brendan Coe	Good evening	tomorrow	guests	toilet
Jack Cruise	Hello	the day after tomorrow	children	drain
Kate Morgan	Hi!	next Thursday	….	fans/radiators

2 A questionnaire: what to say to customers

Customers expect polite and professional behaviour as they are paying for a service. Choose what you think is the best answer in each situation.

1	**A customer asks you a question that you can't answer.**
A	I don't know.
B	I can't help you there.
C	I'll find out and get back to you.
2	**A customer asks you a question about something that you aren't responsible for.**
A	I don't know much about that.
B	Let me pass you over to a colleague who can help you with that.
C	Sorry, that's not my area.
3	**A customer telephones you when you have no time to talk.**
A	Call back later, please.
B	Can I call you back in a few minutes, please?
C	Sorry, I've got no time to speak to you just now.
4	**A customer makes a complaint about a job that has been done badly.**
A	It isn't my fault.
B	What do you want me to do about it?
C	I'm sorry there's been a problem. I'll find out how we can solve it.
5	**A customer asks you to do something just before your working day ends.**
A	I'm finishing work now. Would it be okay if I did it first thing tomorrow morning?
B	Sorry. I've got to go.
C	I can't help you now. I was just leaving.

3 A checklist for good customer service

a) Match the sentence halves to make complete sentences about good customer service.

1 Always be polite to customers,
2 Be punctual and arrive
3 Try to make a good impression by being smart and tidy
4 If a customer complains, listen carefully
5 Try to complete work within the time
6 If a job is going to cost the customer more than the price you originally quoted,
7 Give customers the feeling that you are interested in them
8 Smile and make eye

A contact with customers to make a friendly impression.
B tell him/her about the price increase as soon as you can.
C to what they have to say and then tell them you are sorry there has been a problem.
D even if they are rude to you.
E you have told the customer it will be completed.
F by dealing with one customer at a time.
G at the time you agreed with the customer.
H in appearance and by keeping the company van clean.

b) Discuss with a partner. Which three items do you think are the most important?

IT'S YOUR TURN

With a partner, discuss how you deal with customers in your company. Are you given guidelines describing how you should talk to customers? Create your own guidelines explaining how best to treat customers using language covered in the unit and your own ideas.

3 Dealing with colleagues

This unit is about:
- politeness
- communication

Competences:
- getting on with colleagues
- talking to colleagues
- being polite

1 Getting on with colleagues

It is difficult to go to work every day if you have problems with colleagues. Here are some tips on how to get on with them.

★ a) Form correct sentences by linking the numbers 1–8 and the letters A–H.

1 Always be polite
2 It is a good idea to offer to help your colleagues
3 You will not always like all your colleagues but
4 Don't talk about colleagues
5 Always ask before
6 Don't forget to return the tools
7 Don't interrupt colleagues
8 Never take food from the fridge

A when they are not there.
B when they are talking to each other.
C you borrow a colleague's things, e.g. his/her pipe wrench.
D you have borrowed.
E when they have a problem or too much work.
F that does not belong to you.
G to your colleagues.
H you should still treat them with respect.

★★ b) Read what Bob, another trainee, says he did and tell him what he did wrong.

1 I don't know why Tom was so angry when I ate that sandwich I found in the fridge.

____

2 I know Carsten was talking to Mary but I couldn't understand why he did not answer when I asked him where we keep the gas bottles for the propane torches.

____

3 Okay, so I did not return the pipe cutter I borrowed from Billy last month. Is that why he is angry with me?

____

4 I told Mark that I saw Thomas at the cinema last Friday when he phoned in sick that day.

____

5 I don't like Colin, and yesterday I told him I thought he should make sure he knows what he is talking about before he starts to speak.

____

★★★ c) Bob thinks his supervisor is not treating him fairly. What tips would you give Bob?

Bob's problems
I never seem to be able to talk to him.
He never listens.
I don't think he is interested in my problems.
He always criticizes me.
He never has a good word to say to me.

2 Talking to colleagues politely

There are polite ways of saying something and impolite ways of saying something.

★ a) Match the polite phrases from the box to the impolite phrases below.

> I wonder if you could stop doing that. • Could you talk a bit quieter, please? • I'm afraid I'm quite busy at the moment. • Could you give me that, please? • Could you possibly come back later? • Would you mind waiting?

1 Be quiet!

____

2 Just wait here.

____

3 Come back later.

____

4 Give me that.

____

5 Stop that.

____

6 I have no time.

____

★★ b) The following statements are taken from conversations among colleagues. Mark polite statements with a "p" and impolite statements with an "i". Underline words that express politeness.

1 I can't hear what John is saying because you are making so much noise. ☐
2 I'm afraid I have to refuse your request at the moment. ☐
3 Could you give me that pipe cutter, please? ☐
4 Give me that pipe wrench. ☐
5 I have no time for you at the moment. Can't you see I'm talking to Barbara? ☐
6 I'm afraid I can't help you right now. Could you come back later? ☐
7 No, I can't help you. Can't you see I'm busy? ☐
8 Could you possibly speak a bit quieter, please? I'm talking to a customer. ☐

★★★ c) Now write polite versions of the four impolite sentences above.

____

____

____

____

IT'S YOUR TURN

Imagine you are a supervisor and a new trainee has just started in your company. Make a list of the five most important "dos and "don'ts" to help the trainee to get on with his/her colleagues.
Compare your list with a partner's and then make one list from the two lists.

4 Developments in the plumbing trade

This unit is about:
- video pipe inspection
- a modern pipe repair system

Competences:
- completing a text about plumbing equipment
- understanding a text
- listening to a recording about pipe repair

4 1

Modern technology in the field of plumbing

Here is some of the modern plumbing equipment that often requires special training to operate it.

a) First, try to complete the text with words from the list. Then listen to the recording to check your answers and fill in any gaps.

camera • clean • cleaning • damage • foundation • inside • leaks • monitor • outside • powerful • remove • water

So-called "robotic pipe inspection crawlers" are used for video pipe inspection and have a very small ____________[1] fixed on long, flexible cables. They are used for drain ____________[2], leak detection, sewer and water pipe repairs. The camera sends the information to a video ____________[3], and plumbers can see what the condition of the ____________[4] of a pipe is and then they can decide what the best way is to repair the pipe.

Pressure sensor leak detection devices are used to detect ____________[5], and are very useful for finding leaks that occur in a building ____________[6]. With this equipment, plumbers can reduce the amount of ____________[7] they will have to do to repair the leak.

Hydro-jetters use motors to put ____________[8] under extreme pressure and heat, and then they blast the water out through a hose to ____________[9] the inside of pipes. A hydro-jetter is one of the most effective ways of clearing ____________[10] and inside blocked pipes.

Motorized drain snakes are more ____________[11] than normal drain snakes. Motorized drain snakes reach the toughest blockages and either ____________[12] them or break them up.

★ b) Find the words in the text that match these definitions.

1 something that moves very slowly ____________
2 can easily be bent ____________
3 water coming out where it is not supposed to ____________
4 the place where dirty water goes ____________
5 what state something is in ____________
6 the part on which a building is built ____________

★★ c) Read the text again and say whether these statements are true (T) or false (F).

	T	F
1 A camera on the top of a robotic pipe inspection crawler sends pictures to the plumber's mobile.		
2 You can fix a leak with a robotic pipe inspection crawler.		
3 Plumbers can look into the pipes with a robotic pipe inspection crawler.		
4 Plumbers sometimes do some damage when they use a pressure sensor leak detection device.		
5 Hydro-jetters blast the water through the hose to clean it.		
6 Hydro-jetters use water under extreme pressure.		
7 If you have a really blocked pipe, you can use a snake.		

★★★ d) Correct the false statements.

5 2 Modern pipe repair: using a new epoxy lining system

A plumbing service technician is talking to a customer about pipe lining and pipe repairs.

a) Listen to their dialogue and fill in the missing words.

Well, Mrs Dickson, it depends on the condition of the pipe and its ____________[1]. What we do is we send this robotic pipe ____________[2] crawler into the pipes and then we can see the state of them from the inside. If we find a leak, then, either a ____________[3] epoxy or pipe lining fabric soaked in epoxy can be used to repair it. We can reline both the ____________[4] pipes and ____________[5] pipes. Because we don't have to first dig up the pipes and remove a lot of ____________[6] in order to inspect them, we can reline your water pipes for you really quickly. And when I say quickly, I mean we can line your pipes in one day without any mess while saving you a lot of money. Did you know that to ____________[7] pipes is 50% to 75% more expensive than to line them? And there's no ____________[8] to buildings or surrounding areas such as roads, ____________[9] or gardens.

The pipe lining we use provides a safe, non-toxic, ____________[10] barrier between the pipe and the water. So, you needn't worry about ____________[11] elements getting into your water. This can't happen. And this new lining is very ____________[12]. We estimate that it will last a hundred years. Our spray-on system offers you the most advanced epoxy pipe lining that is ____________[13] today. We have used this method successfully for many hundreds of kilometres of water pipes worldwide.

If you like, I could come round at 2 o'clock today, have a look at the pipe situation at your house and give you an ____________[14] of what the relining process would cost you.

mauritius images/Lightworks Media/Alamy

Tip

Epoxy is best known as a type of glue that is far superior to most ordinary paste-style glues. Epoxy resin can be toughened with the addition of thermoplastic polymers.

b) Read the text again and write down all the numbers and say what they mean.

Example: *one = One day is the length of time it takes to reline pipes.*

IT'S YOUR TURN

You are very interested in modern equipment. Either research one of the pieces of modern equipment mentioned above, particularly the special training needed and report back to the class.
Or report on any modern equipment that you use in your job.

Extra practice: Units 1–4

1 Choosing a career in plumbing

You want to find out why your partner wants to become a plumber.

★ a) Work with a partner and practise the following dialogue.

Student A	Student B
Ask your partner why he/she wants to become a plumber.	
	Tell your partner your parents told you to become a plumber. Ask your partner for his/her reasons.
Tell your partner you have always been interested in plumbing. Ask your partner how he/she found a job.	
	Tell your partner you wrote ten letters of application, went for four interviews and got a job. Ask your partner how he/she found a job.
Tell your partner that your uncle works as a plumber and he got you the job. Ask your partner if he/she has been out with the supervisor on a job yet.	
	Tell your partner that you went to a house yesterday to repair a tap. Ask your partner if he/she has been out with the supervisor on a job.
Tell your partner that you are going out tomorrow and it is also a private house.	
	Remind your partner that he/she must be polite to the customer.

★★ b) Change the underlined parts of the dialogue to make your dialogue different. Present your changed dialogue to the class.

2 Plumbers and their jobs

Here are some jobs that plumbers have done. You want to know more about them.

a) Use the verbs in the box to complete the sentences.

deal with • install • operate • repair • service • unblock

1 Yesterday I had to ______________ new water pipes in an old house that was being renovated.
2 The most difficult job I have done so far is to ______________ the drainage piping in the kitchen of a young family. They had poured so much old fat down the sink.
3 Why do many customers wait until the cold weather before they phone a heating engineer to ______________ their central heating?
4 I know that I have to be very polite when I ______________ customers, even unfriendly ones.
5 This week we have to ______________ the heating and cooling system that another firm installed in a big office block in town last month!
6 Only a plumber with special training can ______________ robotic pipe inspection crawlers.

b) Now get together in groups of three. Make a poster about all the different tasks a plumber has to do. Underline all the tasks you have already done. Present your poster to the class.

3 Dealing with customers

Janet is visiting a new customer. His hot water boiler isn't working.

★ a) Fill in the gaps with what Janet says to the customer.

> Could you turn the volume down a little? • Don't you want to answer it? • Good morning, Mr Baxter. My name is Janet. • Look, these parts are worn out. • We are never late for an appointment.

Janet: ______________________________[1] You asked us to repair your hot water boiler.

Mr Baxter: Yes, that's right. Good morning. You are very punctual. My appointment was for 10 o'clock and it is exactly 10 o'clock!

Janet: ______________________________[2] Now, where is this hot water boiler?

Mr Baxter: Let's go into the kitchen and I'll show you.

Janet: What a lovely kitchen. Did you design it yourself?

Mr Baxter: Yes, I did. I'm glad you like it.

Janet: I'm sorry, Mr Baxter, what did you say? The radio is a little loud. ______________________________[3]

Mr Baxter: Oh, sorry. Is that better?

Janet: That's fine, thanks.

...

Janet: Ah, yes. I think I know what the problem is. ______________________________[4] I will need to order new parts.

Mr Baxter: Will the repair job be expensive?

Janet: I don't know exactly. It depends how expensive the replacement parts are. I will find out and let you know as soon as possible. OK?

Mr Baxter: Thank you. Oh, is that your mobile phone? ______________________________[5]

Janet: No, it will just be my boyfriend. He can leave a message on voicemail and I'll ring him back later.

★★ b) With a partner, make a new dialogue and practise it together.

▶▶▶ ICH KANN

		Seite/Übung
☐	Ich kann auf einfache Art Gegenstände, Geräte und deren Funktion beschreiben. A2	11/ It's your turn
☐	Ich kann an einfachen Gesprächen teilnehmen (z. B. Ausbildung). Ich kann dabei auch zeigen, dass ich mich für das interessiere, was mir jemand sagt. A2	5/It's your turn; 7/It's your turn
☐	Ich kann eine Rolle in einem Dialog oder einer Simulation übernehmen (z. B. Berufssituation). A2	4/1; 6/1c
☐	Ich kann kurzen und deutlich gesprochenen Texten und Tonträgern die wesentlichen Informationen entnehmen, wenn es um vorhersehbare alltägliche oder berufsspezifisch vertraute Dinge geht. A2	4/2; 10/1
☐	Ich kann einfachen Alltagstexten, die sich auf vertraute Situationen beziehen, die wichtigsten Informationen entnehmen (z. B. berufliche Alltagsbeschreibungen). A2	10/1
☐	Ich kann Listen schreiben (z. B. Produktlisten) und einfache Formulare ausfüllen. A1	9/ It's your turn

5 Plumbing systems

This unit is about:
- water supply systems
- drainage and ventilation systems
- gas systems

Competences:
- labelling a diagram
- listening and note taking
- discussing gas systems

A typical plumbing system has got three parts – water supply, waste drainage and ventilation. These three parts aren't separate – they are connected together and form an interdependent system. Water usually comes from a public or private water mains supply pipe under the street, but it can also come from a well or a spring. This supply system brings drinking water under pressure from the mains to the house through a large supply pipe. From here it goes to all the fixtures and appliances inside the house that use water.

1 The water supply system

The water supply system is subdivided into a cold-water system and a hot-water system.

★ a) Study the diagram and colour the two systems blue and red.

blue = cold-water system
red = hot-water system

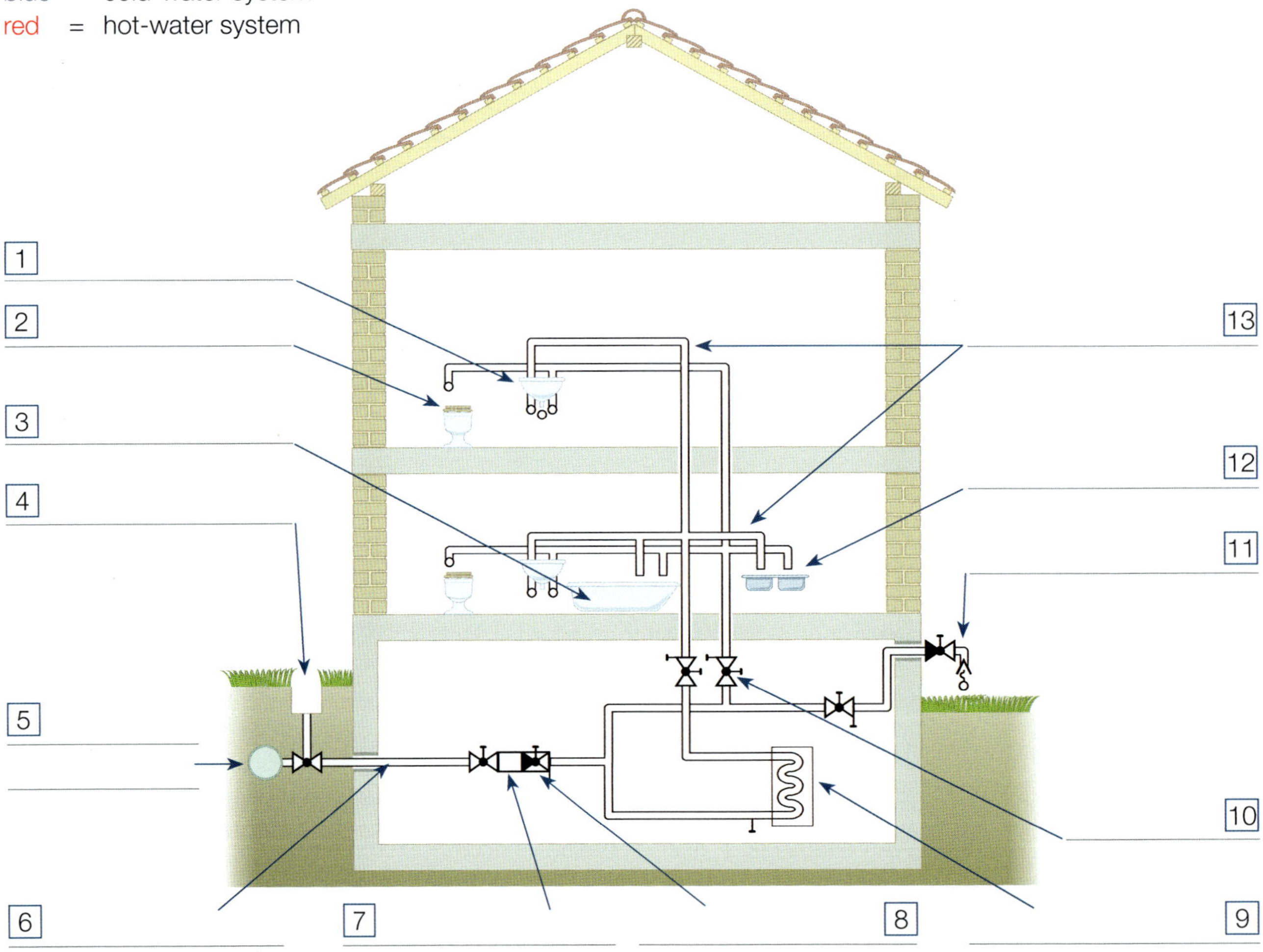

★★ b) Identify the following pipework and appliances and label the diagram with these words.

bathtub • branch pipes • check valve • gate valve • hose bib • kitchen sink • main supply pipe • shut-off valve • supply pipe • toilet • washbasin • water heater • water meter

2 The drainage and ventilation systems

A ventilation system allows sewer gases from the soil stack to escape into the air. It also maintains the correct low pressure inside the drainpipes. The drainage system carries used water and waste out of the house.
Identify the following pipework and label the diagram with these words.

branch drain • cleanout • house drain • main drain • main vent stack • soil stack • trap

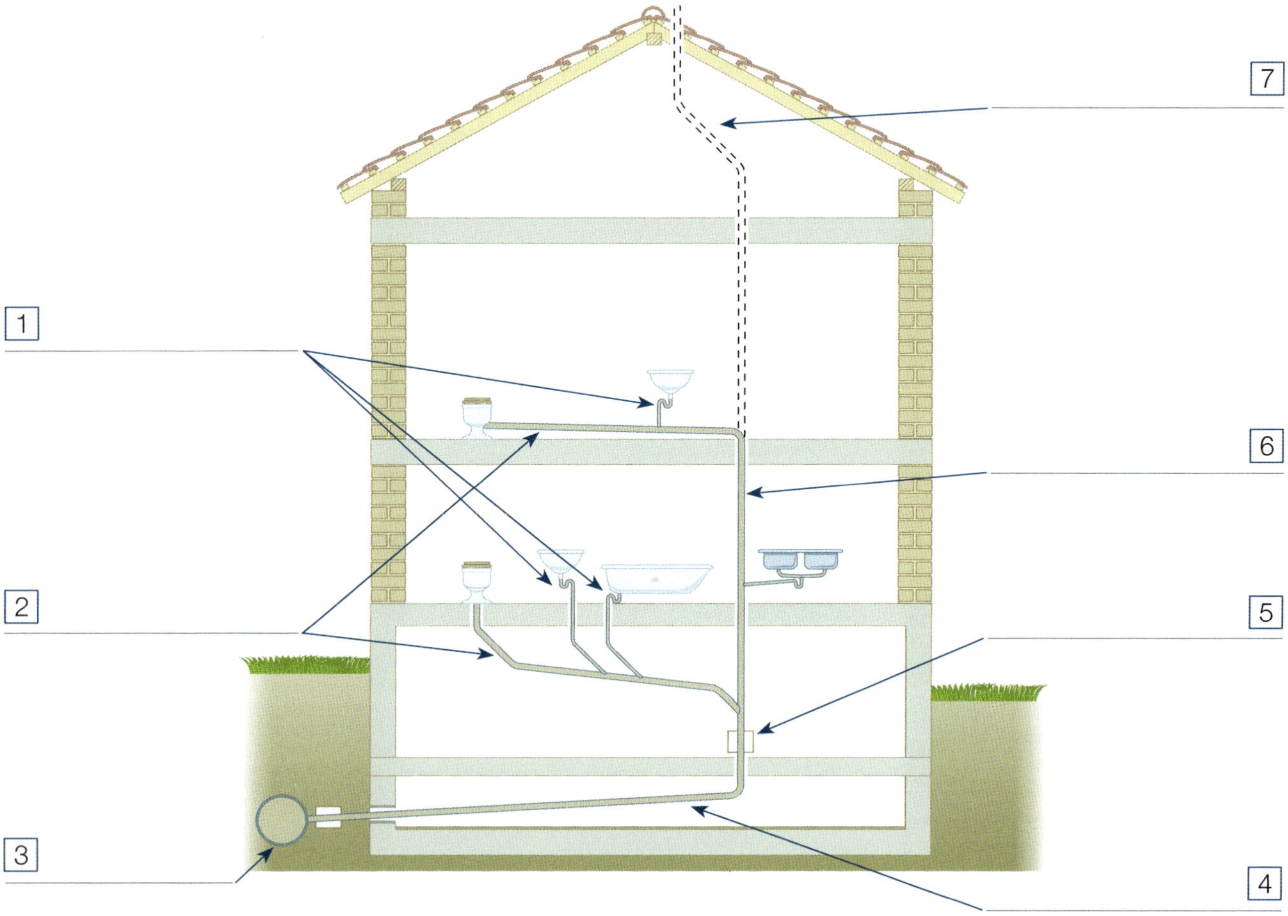

6

3 Gas systems

You want to know more about gas systems. Look at the questions below. Then listen to the recording and take notes. Use your notes to answer these questions in complete sentences.

1 What is gas usually used for in homes?

2 What device measures the amount of gas used?

3 How is propane gas usually stored?

4 How is natural gas supplied to a house?

IT'S YOUR TURN

You want to know how the water system works in a house that is not connected to the public drainage system. Find out how a house gets its water from a well or a spring and report to your partner.
Or explain how a cesspool or a septic tank functions.

6 Plumbing tools

This unit is about:
- a plumber's tools
- an emergency plumbing kit

Competences:
- labelling tools
- identifying sounds of tools
- identifying descriptions of tools

7 🎧 **1 A plumber's toolbox**

To do any job well, you need the right tools and materials – and plumbing is no exception.

★ a) Label the pictures with the words from the box.

adjustable spanner • basin wrench • blunt-jaw pipe wrench • C-clamp • electric drill • files • hacksaw • pipe cutter • pipe vice • pipe wrench • plunger • screwdrivers • spanner • spirit level • wire brush

1 

______

2 

______

3 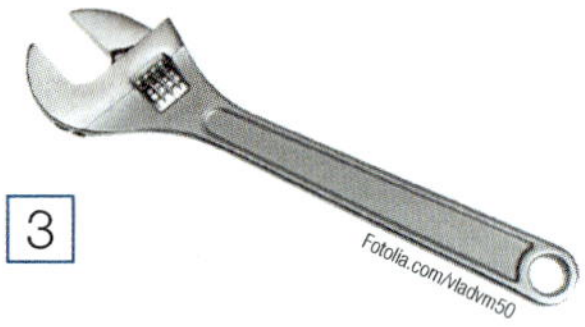

______

4 
______

5 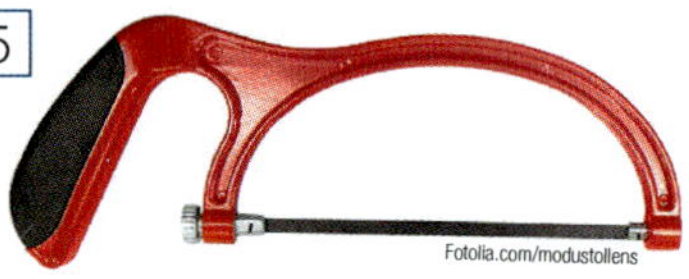

______

6 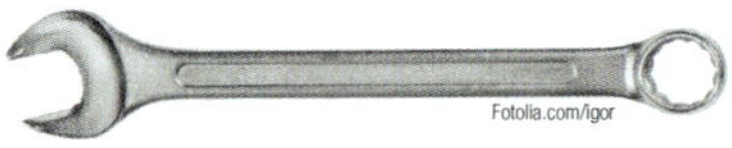

______

7 

______

8 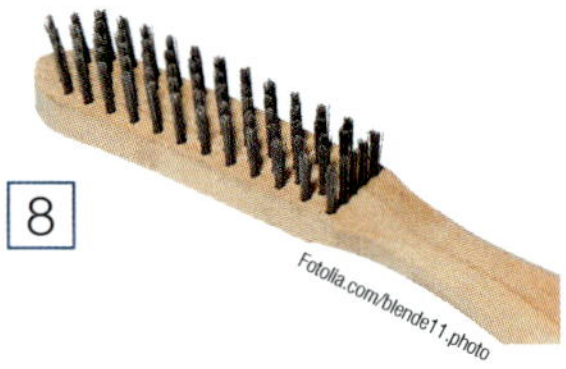

______

9 

______

10 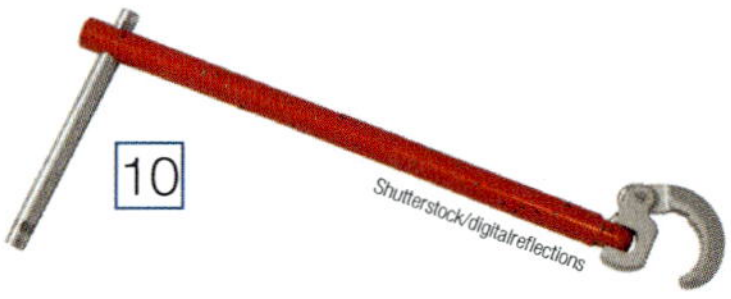

______

11 

______

12 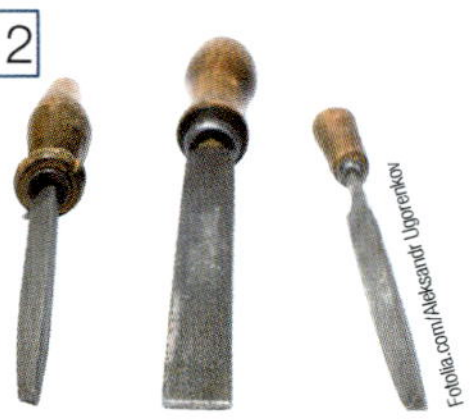

______

13 

______

14 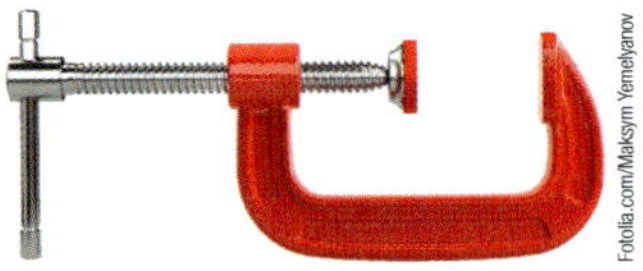

______

15 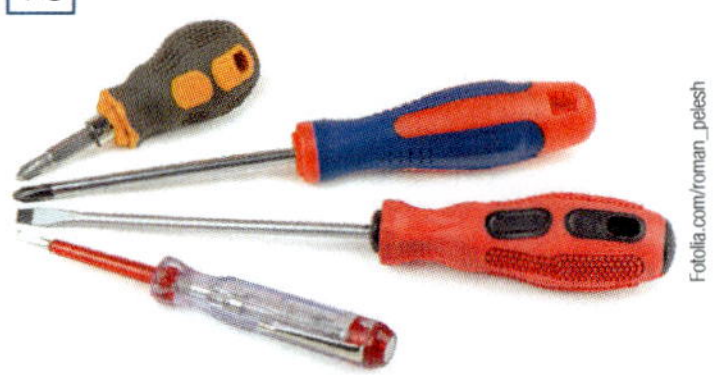

______

★★ b) Listen to these four sounds and name the tools.

1 ______ 3 ______

2 ______ 4 ______

2 Basic tools

Which tools fit these descriptions?

1 It has jaws with teeth so that it can grip a threaded pipe. It's used to tighten fittings on threaded pipes. You can use two of them if you haven't got a pipe vice. ______

2 There are small teeth on the blade and you use it to cut metals. ______

3 It clamps a pipe so that you can cut it and thread it. ______

4 It allows you to reach nuts behind sinks and in other places that are hard to reach. ______

5 It has smooth jaws to fit nuts and square or hexagonal fittings of different sizes. ______

6 This tool is the quickest and easiest way to cut pipes. ______

7 You need this tool for tightening or loosening screws. ______

8 This tool can clear blocked washbasins and toilets. ______

9 It can fix a chromium-plated tap to a washbasin without scratching it. ______

10 It shows you if something is level. ______

3 An emergency plumbing kit

The following materials are essential in an emergency plumbing kit.

★ a) Label the photographs with the terms below.

duct tape • electrical tape • hose clamps • O-rings • packing tape • screws • tap washers

1

______

2

______

3

______

4

______

5
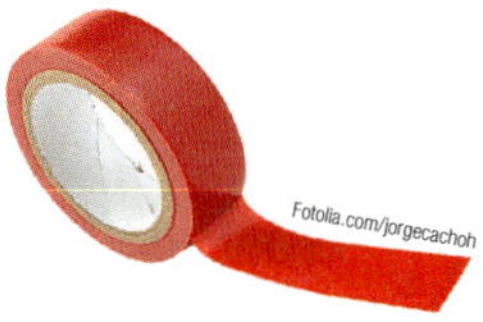

______

6

______

7

______

★★ b) Look at the tools in exercise 1 and the materials above. Choose ten things that you think should be in an emergency plumbing kit.

1 ______ 6 ______
2 ______ 7 ______
3 ______ 8 ______
4 ______ 9 ______
5 ______ 10 ______

IT'S YOUR TURN

Each company uses different tools. With a partner, discuss which tools you use at work. Make a list of the tools you needed last week and then compare your list to your partner's list.

7 Safety at work

This unit is about:
- safety equipment
- safety signs
- safety rules

Competences:
- talking about safety equipment
- describing safety signs
- understanding safety information

1 Personal safety

Health and safety is very important and so all plumbers must understand typical safety signs.

★ a) Link the five safety items to the illustration.

Protective headgear:	safety helmet	☐
Eye protection:	goggles	☐
Hearing protection:	ear protectors	☐
Body protection:	a pair of overalls	☐
Protection from falling:	scaffolding	☐

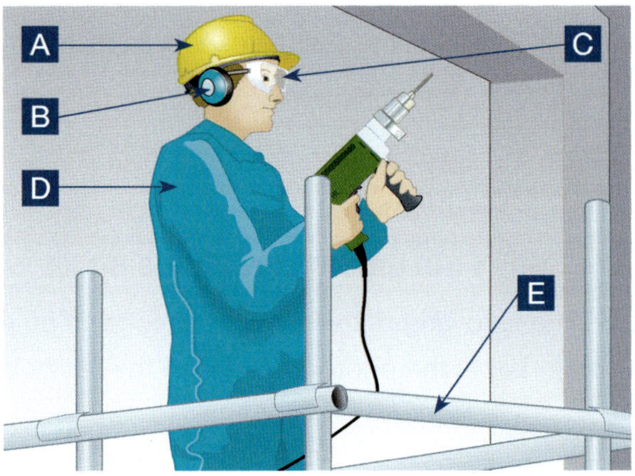

★★ b) What safety equipment is it?

1 It protects your head. ______________________

2 They protect your ears. ______________________

3 They protect your eyes. ______________________

4 You need this if you're working on a high wall. ______________________

5 It's in one piece and you wear it over your other clothes. It protects both you and your clothes.

____

2 Safety signs

★ a) What do these symbols mean? The words and phrases in the boxes may help you.

protect • use • wear	ear protectors • goggles • a respirator • safety boots • safety gloves • a safety helmet	your eyes • your feet • your hands • your head • your hearing

1 ______________________

2 ______________________

3 ______________________

4 ______________________

5 ______________________

6 ______________________

★★ b) Think of a situation where you would need to wear these pieces of safety equipment and make sentences.

3 Safety rules for angle grinders

You must always read the safety instructions before you use a new tool for the first time. Complete the sentences with a term from the box.

disc guard • discs • earplugs • goggles • off • overalls • power cable • screwdriver • side handle

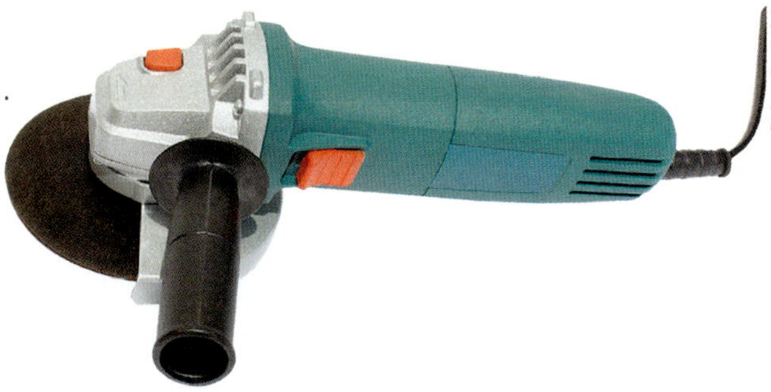

Fotolia.com/Angelika Smile

1 Loose clothing can be caught in moving parts, so always wear ______________________.
2 Wear ______________________ to protect your eyes from flying parts.
3 Always use ______________________ because an angle grinder is very noisy.
4 Make sure the switch is in the ______________________ position before plugging the machine into an electrical outlet.
5 Keep the ______________________ well away from the grinding disc.
6 Always disconnect the grinder from the electric outlet before changing ______________________.
7 Always use the ______________________ when using the machine.
8 Before you use a grinding or a cutting disc make sure it isn't damaged by tapping it gently with a ______________________. If the wheel sounds dead, don't use it!
9 Never use the grinder without the ______________________.

8

4 Safety information for a pipe-threading machine

Listen to the recording of a supervisor telling his apprentices about safety information and then complete the sentences with a verb from the box.

block • keep • operate • reach • stop • use (2×) • wear • work

Shutterstock/Taechit Tanantornanutra

1 Never ______________________ in loose clothing.
2 Always ______________________ a pair of overalls.
3 Never ______________________ gloves.
4 Always ______________________ the footswitch.
5 Never ______________________ the footswitch.
6 Always ______________________ the machine from the side where the footswitch is.
7 Never ______________________ across the machine.
8 Always ______________________ your hands well away from the rotating pipe.
9 Always ______________________ the machine when screwing fittings onto it.

IT'S YOUR TURN

Your boss has explained to you how important safety is at work. Make a list of the five most important safety rules that you know or think are the most important from this unit and say what could happen if these rules were broken.

8 Materials used in plumbing

This unit is about:
- materials used in plumbing
- fittings

Competences:
- labelling types of steel
- listening to a conversation about plumbing materials
- identifying fittings

1 Steel

Cast iron is easy to shape, but it breaks easily and it rusts. The three types of steel in the list below don't break easily and they don't rust. This makes them popular plumbing materials for bathrooms and kitchens. Look at the photos. What are these things and what are they made of? Use words from both boxes.

chrome-plated steel • galvanized steel • stainless steel

adjustable pipe wrench • kitchen sink • mixer taps • pipe fitting • public toilet washbasin • screw • shower head • towel rail • water pipe • water tray

1

Robert Nadolny, Berlin

2
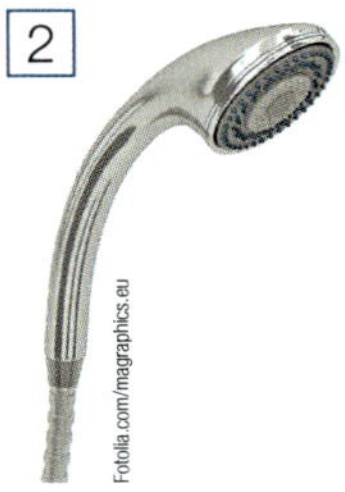
Fotolia.com/magraphics.eu

3
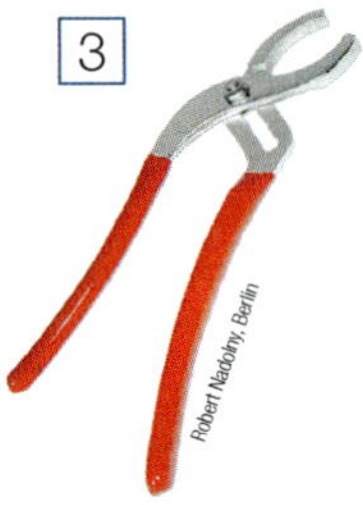
Robert Nadolny, Berlin

4

Shutterstock/severike

5
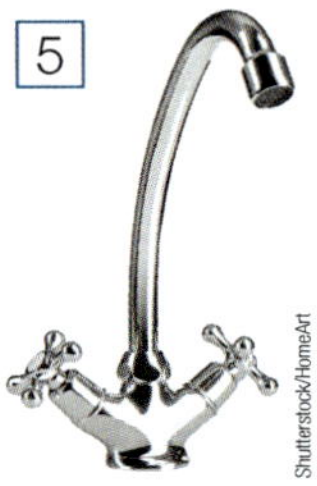
Shutterstock/HomeArt

6

Fotolia.com/fablok

7
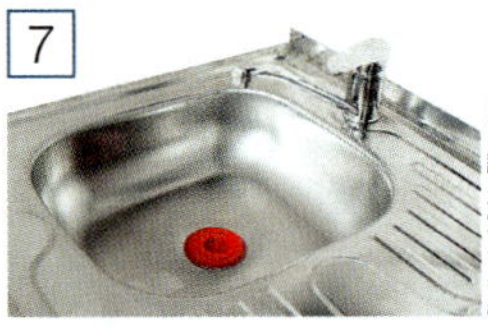
Shutterstock/Iakov Filimonov

8

Fotolia.com/PVMil

9
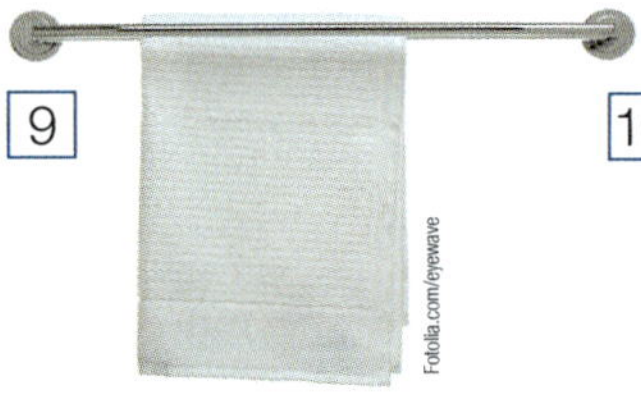
Fotolia.com/eyewave

10

Robert Nadolny, Berlin

2 Copper, bronze, brass and plastic

9

For your work you need to know more about why certain materials are good for plumbing.

★ a) Listen to Janet's supervisor explaining plumbing materials. Which words from the box go with the materials in the table?

cheap • expensive • flexible • hard • long-lasting • rust-free • soft • water-resistant • weak

Copper	Brass/Bronze	Plastic

★★ b) Work with a partner. Listen to the recording again, take notes and write down the other information about copper, bronze, brass and plastic that you hear.

3 Cast iron

Iron is cheap and it's a perfect material for casting. Casting is a process where the molten iron is poured into a mould. This is a cheap way of producing things with simple shapes. These products are very robust and last for a long time. These things are made of cast iron. What are they? Label the photos with words from the box.

drainage grate • drainage pipe • manhole cover • pipe connector • pump housing • radiator

1 

____

2 

____

3 

____

4 

____

5 

____

6 

____

4 Fittings

Fitting pipes is typical plumber's job. Read the text and label the illustrations with the words in blue.

To fit things such as a stopcock, a tap, or a U-bend trap you will need an adjustable pipe wrench (water pump pliers). To join a pipe fitting onto a pipe with a thread you will need two pipe wrenches – one to hold the pipe and the other to turn the fitting.

1 

____

2 

____

3 

____

4 

____

5 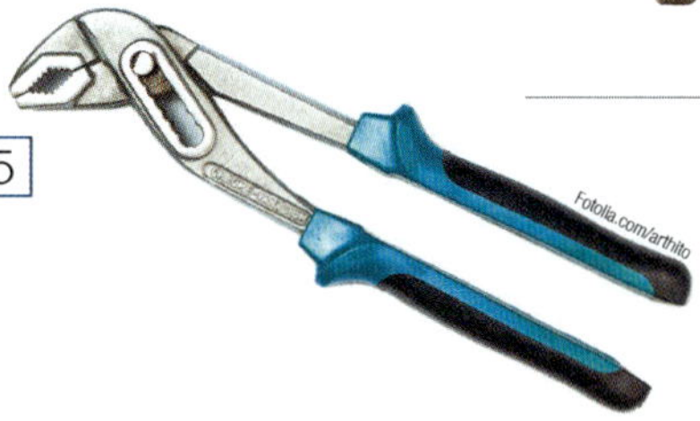

____

6 

____

IT'S YOUR TURN

Make a mind map of the most important materials used in plumbing. Add the German translations and words you can use to describe their properties adding any other information you know about the materials. Then compare your mind map with a partner.

Extra practice: Units 5–8

10 1 Safety first

Richard is a pipefitter who often forgets to follow the company's safety procedures. Mr Coppering, his supervisor, calls him in for a chat.

a) **Try and complete the dialogue. Then listen to the recording, check your results and fill in any gaps.**

Mr Coppering: So, Richard, this is the third time this week that you have forgotten your ________________ [1] helmet at your workplace.

Richard: Yes, I know but it is always so hot in summer.

Mr Coppering: That is true but what about your ear ________________ [2]? Where are they? Why don't you wear them?

Richard: But then I won't hear you when you call me.

Mr Coppering: But you should always use them because it is very noisy when you work with the pipe-threading machine. And how often have I told you not to use a ladder when you are installing pipes on a high wall? What should you use?

Richard: I know, I know. I should use the ________________ [3], but a ladder is quicker to put in place.

Mr Coppering: But that's not the point, is it? And what's more, I saw that you were not wearing your ________________ [4] this morning. You must protect your feet. If something heavy falls on your feet, you could get seriously hurt. You know that, don't you?

Richard: Yes, I do. Sorry.

Mr Coppering: Oh, and I heard that last week you did not wear safety ________________ [5] on your hands when you were using the grinder.

Richard: But I was very careful.

Mr Coppering: I really don't think you are careful enough. Let me ask you: what should you not wear when you use a grinder, what should you wear and why?

Richard: Jeans? A jacket? I don't know.

Mr Coppering: You should never wear ________________ [6]. They can be caught in the moving parts. You should always wear ________________ [7]. Please take this leaflet, read it carefully and learn the safety ________________ [8] that we all need to take when working.

b) **Now work with a partner and make up a similar dialogue which you practise with him/her.**

2 Ventilation and drainage systems

a) **A British trainee plumber has started at your place of work and has to write a report. He is not sure of the difference between drainage and ventilation systems.**
Help the trainee identify the correct system.

1 This system maintains the correct low pressure inside the drainpipes. ________________

2 This system carries used water away from the house. ________________

3 This system lets sewer gases escape into the air. ________________

4 This system carries waste out of the house. ________________

b) **Now prepare a short presentation on drainage and ventilation systems and then give it to your class.**

3 Materials

You are not sure which material is the best for installations in the house.
Work with a partner and practise the following dialogue.

You	Supervisor
Fragen Sie Ihre/n Ausbilder/in, was der Unterschied zwischen verzinktem Stahl und Gusseisen ist.	Sagen Sie, dass Gusseisen sehr schnell bricht und verzinkter Stahl nicht rostet.
Fragen Sie den/die Ausbilder/in, welches Material im Sanitärbereich das billigste ist.	Sagen Sie, dass Plastik am billigsten ist.
Fragen Sie, warum man nicht immer Plastik verwendet, wenn es preisgünstig ist.	Sagen Sie, dass Plastik nicht immer geeignet ist, da es sehr weich ist.
Fragen Sie, ob Haupthähne aus Gusseisen oder verzinktem Stahl hergestellt werden.	Sagen Sie, dass Haupthähne entweder aus Messing oder Bronze hergestellt werden. Weisen Sie darauf hin, dass Messing und Bronze viel härter als Kupfer sind und auch länger halten.

▶▶▶ ICH KANN

	Seite/Übung
☐ Ich kann auf einfache Art Gegenstände, Geräte und deren Funktion beschreiben. A2	15/It's your turn; 20/1
☐ Ich kann an einfachen Gesprächen teilnehmen (z. B. Ausbildung). Ich kann dabei auch zeigen, dass ich mich für das interessiere, was mir jemand sagt. A2	17/It's your turn
☐ Ich kann kurzen und deutlich gesprochenen Texten und Tonträgern die wesentlichen Informationen entnehmen, wenn es um vorhersehbare alltägliche oder berufsspezifisch vertraute Dinge geht. A2	15/3
☐ Ich kann Anweisungen, Fragen und Auskünfte in Alltags-, Ausbildungs- und Berufssituationen im Wesentlichen verstehen. A2	19/4
☐ Ich kann Interviews, Berichten und Vorträgen in den wesentlichen Punkten folgen. B1	20/2b
☐ Ich kann einfache Sachtexte verstehen, wenn Bilder oder Grafiken den Inhalt unterstützen. A2	14/1; 15/2
☐ Ich verstehe Arbeitsanleitungen in meiner Ausbildung und am Arbeitsplatz und schriftliche Anmerkungen zu meinen eigenen Arbeiten. A2	21/4
☐ Ich kann Listen schreiben (z. B. Produktlisten) und einfache Formulare ausfüllen. A1	17/ It's your turn
☐ Ich kann einfache beschreibende Texte zu grundlegenden Sachverhalten oder Aufgabenstellungen aus meinem Fachgebiet verfassen. B1	18/2b

9 Joining pipes

This unit is about:
- joining pipes
- soldering
- plastic pipes

Competences:
- linking definitions to tools and illustrations
- listening and note taking
- describing how to join plastic pipes

1 Tools for joining copper pipes

There are several ways to join copper pipes. You can solder them together or you can use compression joints. You can use compression joints with both rigid and flexible pipes.
Match the definitions below with the right tool and the correct illustration.

Definition	Tool
1 The quickest and easiest way to cut pipes.	☐ flux
2 They clean and polish the pipe before soldering it.	☐ propane torch
3 This paste prevents oxidation during the soldering process.	☐ soldering wire
4 You use it to apply the flux to the parts you are going to solder.	☐ pipe cutter
5 This alloy of tin and copper melts and fixes the joint.	☐ brush
6 This gas produces a very hot flame that can melt soldering wire.	☐ steel wool + emery paper

Illustration

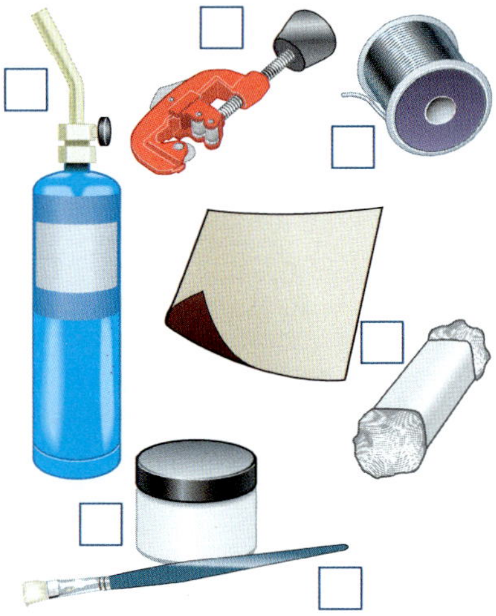

2 Soldering copper pipes

Plumbers often have to solder copper pipes. The pictures showing how to do this are in the correct order, but the text isn't. Match the pictures with the text.

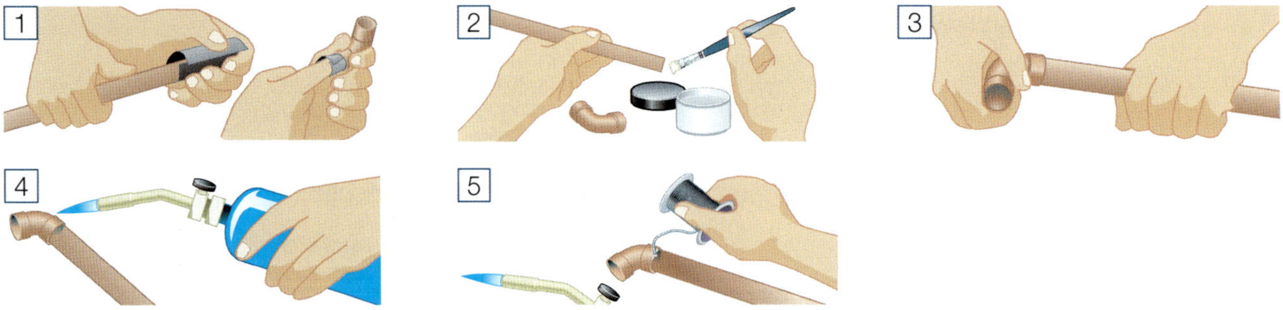

A Push the parts together and twist to distribute the flux evenly. ☐

B Remove the heat and touch the solder to the hot joint where the pipe enters the fitting. The solder will be drawn into the joint. Wipe clean with a rag when it is cool enough. ☐

C After cutting, remove the burrs with the reamer. Polish the outside of the pipe and the inside of the fitting with either emery paper or steel wool. ☐

D Apply a thin layer of flux to the polished areas. ☐

E Heat the fitting evenly with a propane torch until the flux bubbles. ☐

11

3 Plastic pipes

A supervisor is explaining why plumbers sometimes use plastic pipes. Work with a partner. Listen to the recording and take notes. Use your notes to make a list of the advantages and disadvantages of plastic pipes.

Advantages
Disadvantages

Shutterstock / Phovoir

4 Joining plastic pipes

You explain to an English-speaking colleague how to join pipes. Complete these instructions by finding the correct endings to the sentences.

1 Plastic pipes and fittings can be
2 Cut the pipe to the correct size
3 If there are burrs, remove them with
4 Check the pipe carefully – there must be
5 Press the fitting
6 Never lubricate the fitting, the O-ring
7 Place the jaws of the compression pliers in
8 Close the
9 After compression, the jaws of the pliers

A no dirt inside it.
B the guide grooves of the fitting.
C joined by compression.
D firmly into the pipe.
E will open automatically.
F with a pipe cutter.
G a pipe reamer.
H pliers firmly.
I or the end of the pipe.

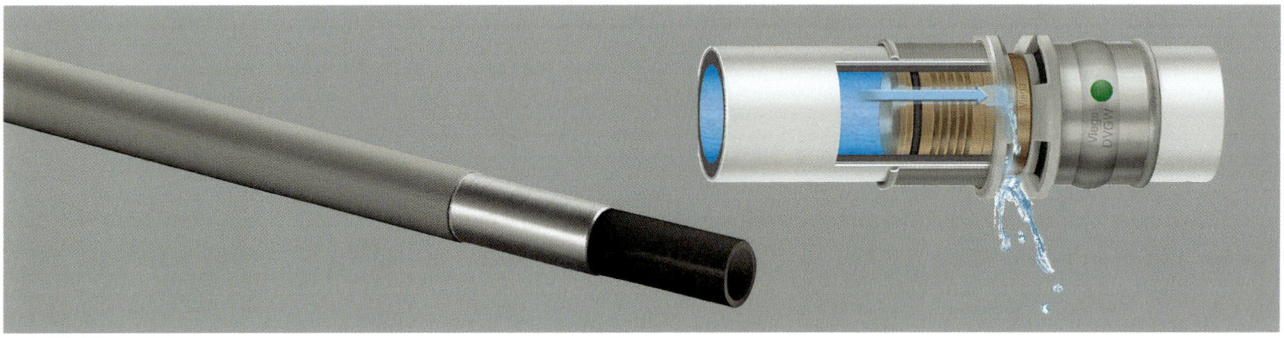

IT'S YOUR TURN

For your work you need to know more about the advantages and disadvantages of copper pipes. Research this with a partner and present your results in a table like the one in exercise 3. Report your findings to the class.

10 Bathroom furnishings

This unit is about:
- bathroom furnishings
- taps

Competences:
- labelling photos
- listening for information
- understanding definitions for bathroom furnishings

1 In a bathroom

Plumbers do a lot of jobs in bathrooms, so they need to know bathroom vocabulary.

★ a) Label the illustrations with the items from the box.

bar of soap • bath/bathtub • bathroom cabinet • bath towel • cistern • hand-held shower • laundry bin • plug • roll of toilet paper • scales • sponge • tap • toilet bowl • toothbrush • towel rail • tube of toothpaste • washbasin

★★ b) Work with a partner. Student A closes his/her book and student B says a word from the box. Student A must say what the object is in German. Now swap roles. Who knew the most words?

12

2 A new bathroom

Ewan and Mandy are looking at a brochure of bathroom furnishings.

★ a) Listen to them planning their new bathroom and complete this table.

salmon | lilac | lime green | pink | sky blue | brushed aluminium | anthracite grey

bathtub	washbasin	fixtures	bathroom cabinet

★★ b) Use the table above and interview a partner about the colours of his/her bathroom at home.

3 Taps

A plumber works with a lot of different types of taps. Label the four different kinds below.

mixer tap • mixer tap for sinks • thermostat-mixer tap • two-grip tap

1

Robert Nadolny, Berlin

____

2

Robert Nadolny, Berlin

____

3

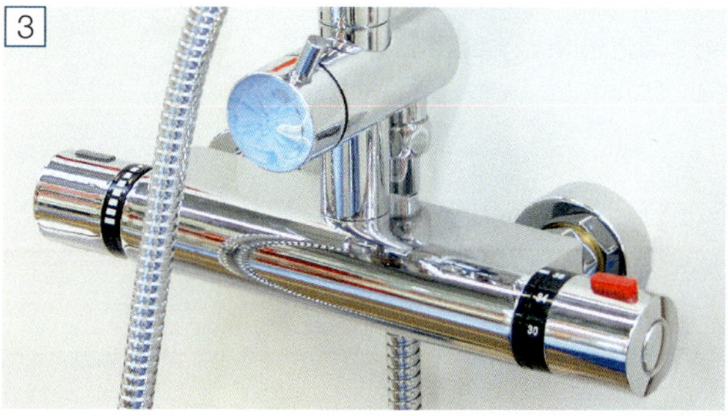

Robert Nadolny, Berlin

____

4

Robert Nadolny, Berlin

____

4 Definitions

★ a) What are they describing? The words are all from the unit.

1 A container of water used for flushing the toilet. In Europe it was invented by the Cistercian monks.

____

2 It turns the water on and off. In America it's called a faucet. ______________________________

3 You fill it up with warm water and then you bath in it. ______________________________

4 It's often part of a bath and it's very useful when you wash your hair. ______________________________

5 It's fixed to the wall and you can wash your hands and face in it – or use it when brushing your teeth.

____

6 It's round and flat and made of rubber. It blocks the outlet of a bath or a handbasin or a sink so that the water stays in and doesn't run out. ______________________________

7 This is where most people keep things such as toothbrushes, razor blades, deodorants and medicines.

____

★★ b) Write five more definitions for bathroom furnishings and give them to a partner to fill in the correct word.

IT'S YOUR TURN

Work with a partner. You have moved into a new flat and you need to renovate the bathroom completely. Discuss what bathroom furnishings and what colour scheme you want. Try to reach an agreement.

11 Toilets and cisterns

This unit is about:
- installing a toilet
- installing a toilet cistern
- fitting a toilet frame

Competences:
- understanding a text
- listening for information
- identifying words

1 Installing a floor-mounted toilet

Installing toilets is a task that plumbers often have to do.

★ a) **Read these instructions and then answer the questions below in complete sentences.**

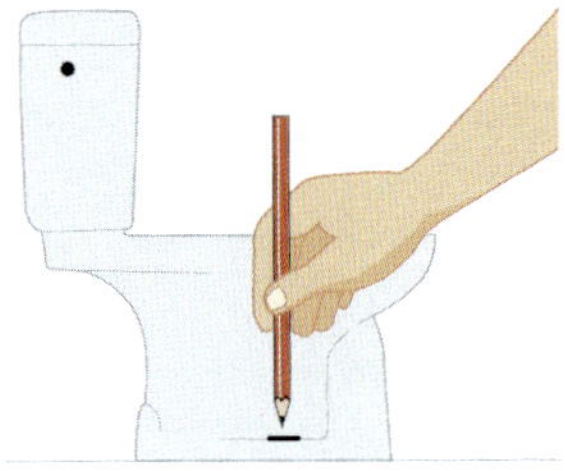

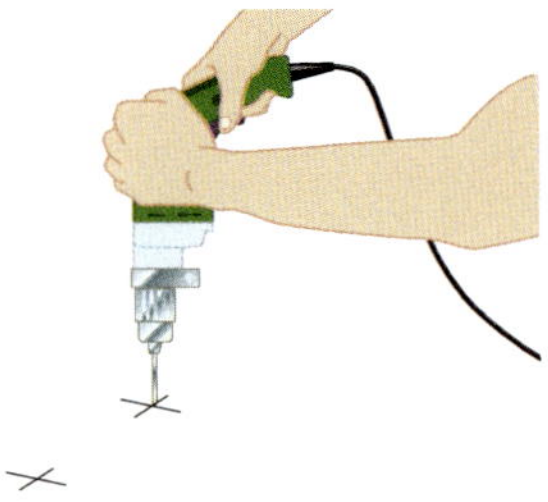

Put the bowl in the correct position. It must fit exactly into the waste pipe. Mark the holes with a pencil or a marker.
Remove the bowl and drill the holes with an electric drill. Use a special drill bit for tiles or concrete. Put plugs into the holes.

Lay the bowl upside down on a bed of newspapers.
The glazed china scratches easily, so do your best to protect it. Put a thick layer of plumber's putty all around the edge of the toilet base.

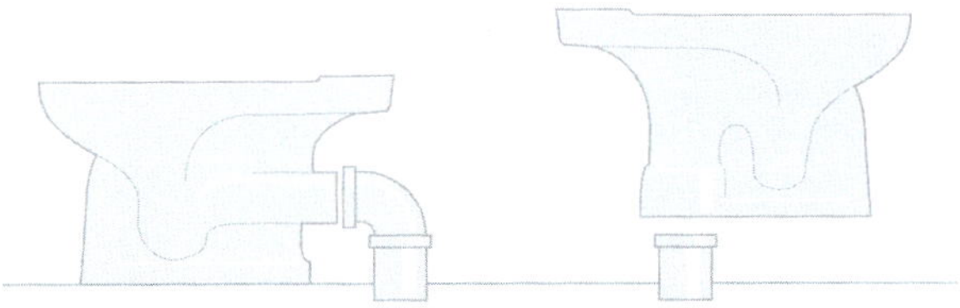

Turn the toilet over and set it smoothly in place in the sleeve of the waste pipe. Make sure the gasket is in the right position inside the waste pipe. Apply a thin layer of gliding paste to the gasket. Press it firmly into the sleeve of the waste outlet pipe.

Use a spirit level to make sure that the toilet bowl is level from front to back and from side to side. If an adjustment is needed, slip thin metal shims under the base of the bowl to get it level and to eliminate any rocking. Put plastic washers on the screws and tighten them with a screwdriver, but not too much. Be very careful or you will crack the bowl!

1 How do you mark the holes?

______

2 What should you do with the toilet bowl before you apply the plumber's putty?

______

3 What should you apply to the gasket inside the sleeve of the waste pipe?

______

4 What do you use to make sure that the toilet bowl is level?

______

5 What will happen if the screws that fix the bowl to the floor are tightened too much?

______

★★ b) **Underline the key words and sum up the text in no more than 70 words.**

2 Mounting a toilet cistern

★ a) What do you know about mounting a toilet cistern?
Label the illustration with words from the box.

cistern • rubber gasket • rubber washers • toilet bowl • wing nuts

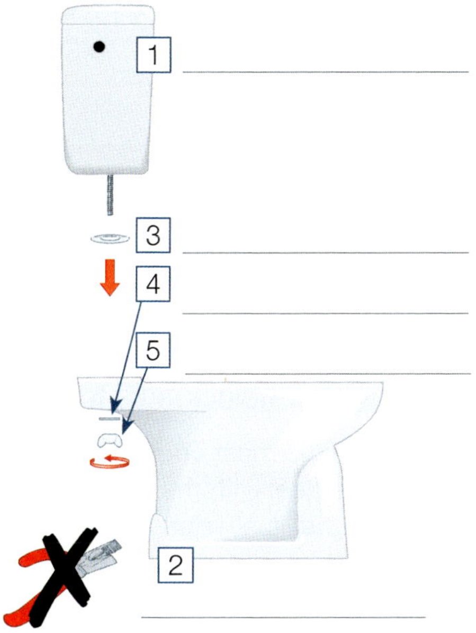

★★ b) Now complete the text with words from the box.

To mount the ______[1] on the ______[2], place the rubber cushion so it lines up with the bolt holes. Push the ______[3] over the cistern's flushing outlet and set the cistern in place. Put ______[4] on the ends of the two bolts and screw on the two ______[5]. Always tighten them by hand. If they are tightened too much, they can crack the toilet bowl.

13

3 Fixing a toilet frame in front of an existing wall

With this system, the plumbing can be hidden behind a newly constructed false wall. Listen to a supervisor explaining how to do this and put the instructions below into the correct order.

Step ☐ Fit the brackets to the frame.
Step ☐ Secure and tighten the wall and feet fixing brackets.
Step ☐ Plug the holes.
Step ☐ Make final adjustments to the level and depth measurements.
Step ☐ Drill holes into the wall for the brackets.
Step ☐ Now fix the locking nuts of the brackets.
Step ☐ Measure and mark the wall taking the final floor levels into account.

4 Quiz

What things from this unit are being described here?

1 It's another word for porcelain. ______
2 A rubber ring used to seal an outlet pipe. ______
3 A small rubber ring that goes over a metal screw. ______
4 A special nut that you can tighten by hand. ______
5 It will stop a screw from falling out of a hole. ______

IT'S YOUR TURN

You read this forum entry and want to help Steve. Write an answer using information from the unit and your own knowledge.

11-24-.., 07:15 PM visiting guest
Toilet installation problems
Hello,
I have nearly completely remodelled my bathroom, but I've just realized that I may have problems with the installation of the toilet. Has anybody got any experience they can share? I would be grateful for any help you can offer.
Thanks,
Steve

12 Blocked fixtures and drains

This unit is about:
- unblocking a toilet, a washbasin
- unclogging a bath drain

Competences:
- labelling an illustration
- listening for information
- writing instructions

1 Unblocking a toilet

One of the most common reasons a customer calls a plumber is a blocked toilet, so it is important to know how to deal with the problem.
Label the illustrations below with the words in blue.

You always know when a toilet is blocked – the bowl fills and won't drain and can even overflow all over the bathroom. The first thing to try is a plunger. This is a rubber cone fixed to a wooden handle. It's easy to use and usually effective with small blockages. The rubber cone fits over the outflow pipe and when the handle is pressed downwards it pushes the water in the pipe against the blockage. By pumping hard, most blockages can be pushed through the U-bend of the trap, which frees the outflow pipe.

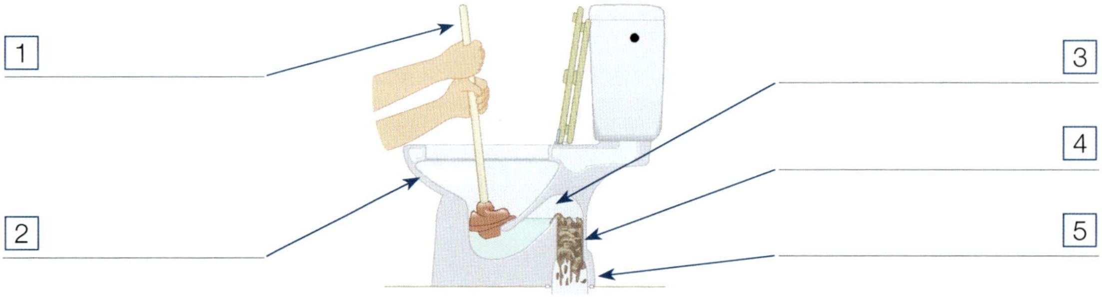

Bigger blockages can't always be freed by a plunger. For these you'll need a flexible auger. This is a metal spiral inside a flexible tubular guide. They are sometimes called 'plumber's snakes'. The tube is pushed into the trap until it meets the obstruction. Turning the handle at the top of the rod rotates the spiral, which drills into the blockage like a corkscrew and loosens it enough to free the outlet pipe.

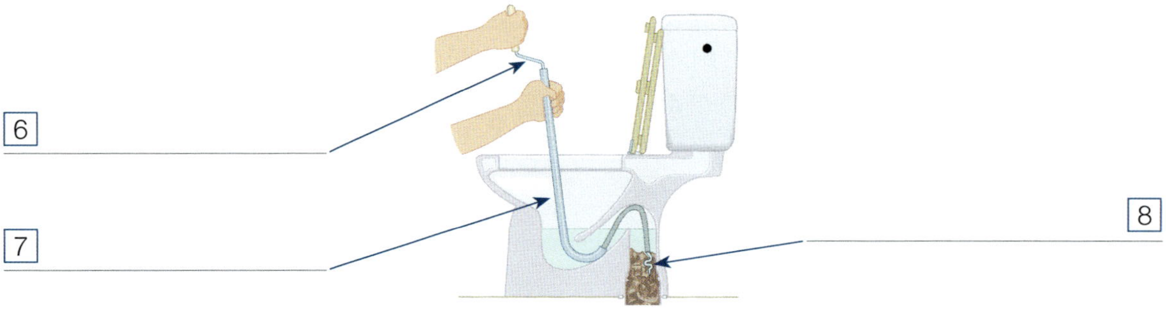

14 🎧

2 Blockages are good for business

Three customers call Bill Sykes, a plumber at Mason Plumbing Services. Listen to the three calls and fill in the details below.

	Name	Problem	Appointment
1			
2			
3			

3 Unblocking a washbasin

Don't use chemical drain cleaners. They are caustic and could burn you if you have to work on the pipes. Chemical cleaners can also form a permanent clog by melting the obstruction into one solid piece before the water can wash it away.

★ a) Look at the pictures and put the instructions describing how to unblock a washbasin into the correct order.

A If the plunger doesn't work and the trap has a cleanout plug, remove it and try to free the clog with a piece of hooked wire.
B Place the plunger over the outlet drain and pump up and down briskly.
C Then plug the overflow hole with a wet rag.
D The final option is to remove the trap and push a flexible auger down the outflow pipe.
E First fill the washbasin half full of water.

1 ☐ ☐ ☐

2 ☐

3 ☐

★★ b) Work with a partner. Write a step-by-step guide for unclogging a bath drain. You can use the texts in exercises 1 and 3 to help you.

How to unclog a bath drain:

___

___

___

___

___

___

___

___

IT'S YOUR TURN

You want to practise making an appointment in English for your customers who only speak English. Work with a partner to write a short dialogue. Partner A is a plumber and Partner B is a customer who calls to make an appointment because of a problem. The language in the box will help you.

(Company name) … speaking. Can I help you? / How can I help you?
I can be with you in half an hour / an hour / … minutes / … Is that okay?
I can come over at … (time) / this morning/afternoon. Would that be convenient?
I have a problem with …
Would it be possible for you to come now? / It's a bit of an emergency. / When would you be able to call by?

Extra practice: Units 9–12

1 A British plumber

A British plumber has come on an exchange trip to your company for four weeks. Your supervisor has asked you to show him around. Answer his questions.

1 Which methods do you use to join copper pipes?
2 Do you use chemical drain cleaners in this country?
3 What kind of bathroom taps do you have in this country?
4 Do you find it easy to deal with customers?
5 Do more households have floor-mounted toilets than toilets that are fixed on the wall?

2 Installing a floor-mounted toilet

A new American trainee has started at your place of work. He thinks he knows how to do things but he often gets things wrong. Help the trainee.

1 **Trainee:** The waste pipe must fit into the toilet bowl.
You: No, ____________________

2 **Trainee:** OK, but I drill the holes into the toilet bowl.
You: No, ____________________

3 **Trainee:** Hmm, I see. Tell me again why I have to lay the bowl upside down.
You: Because ____________________

4 **Trainee:** And where do I put the putty? Around the holes?
You: ____________________

5 **Trainee:** I put the gasket inside the waste pipe, don't I?
You: ____________________

6 **Trainee:** How can I be sure the toilet is level? Do I just sit on it?
You: ____________________

7 **Trainee:** And I need to eliminate any rocking with gliding paste.
You: ____________________

8 **Trainee:** I need to be careful. If I am not careful, I will crack the outlet pipe.
You: ____________________

3 Blocked fixtures

Maria is phoning Harry because she has a problem. Look at these details and, with a partner, write a dialogue.

Harry	Maria
Toilet really blocked?	Bowl fills and won't drain
Used plunger?	Plunger?
Rubber cone fixed to a wooden handle	No
Put it over the outflow pipe	And then? Just wait?
Press it down and pump hard	OK
	Didn't work!
Call plumber	What plumber do?
Plumber's snake	Snake?
Not a real snake – a metal spiral inside a flexible tubular guide	How does that work?
The plumber turns the handle … corkscrew	

4 An article for a trade magazine

Your supervisor has asked you to write an article of about 80–100 words for a trade magazine about the different types of taps and their advantages and disadvantages and where they are best used.

Look at the photos in unit 10 and think about the following:

- easy for children / old people to use
- no risk of the water being too hot because you can pre-set the temperature
- difficult to use if you are handicapped / if you have oily/greasy hands
- usually installed in the kitchen/bathroom / …
- can be used where there is hot and cold water
- can be used where there is only cold water
- easy to clean
- easy to see the temperature of the water

▶▶▶ ICH KANN

		Seite/Übung
☐	Ich kann eine eingeübte, kurze und einfache Präsentation zu einem Thema vortragen, das mir vertraut ist. Ich kann dabei einfache Nachfragen beantworten, wenn ich um Wiederholung bitten kann. A2	25/ It's your turn
☐	Ich kann eine Rolle in einem Dialog oder einer Simulation übernehmen (z. B. Berufssituation). A2	31/ It's your turn
☐	Ich kann ein Gespräch über mir vertraute Themen beginnen, in Gang halten und beenden. B1	27/ It's your turn
☐	Ich kann Anweisungen, Fragen und Auskünfte in Alltags-, Ausbildungs- und Berufssituationen im Wesentlichen verstehen. A2	30/2
☐	Ich kann Interviews, Berichten und Vorträgen in den wesentlichen Punkten folgen. B1	25/3, 26/2
☐	Ich kann einfache Sachtexte verstehen, wenn Bilder oder Grafiken den Inhalt unterstützen. A2	24/2
☐	Ich verstehe Arbeitsanleitungen in meiner Ausbildung und am Arbeitsplatz und schriftliche Anmerkungen zu meinen eigenen Arbeiten. A2	28/1
☐	Ich kann Informationen in verschiedenen längeren Texten oder Textteilen finden, die ich brauche, um eine bestimmte Aufgabe zu lösen. B1	30/1
☐	Ich kann aus vorgegebenen Wörtern einfache Sätze bilden und aufschreiben. A1	25/4
☐	Ich kann mit Hilfe von vorgegebenen Wörtern, Illustrationen und Filmmaterial eine kurze, einfache Geschichte schreiben oder Vorgänge und Abläufe beschreiben. A2	28/1b
☐	Ich kann einfache beschreibende Texte zu grundlegenden Sachverhalten oder Aufgabenstellungen aus meinem Fachgebiet (z. B. *Step-by-step guides*) verfassen. B1	31/3b
☐	Ich kann einfache private Korrespondenz sowie berufliche Standardkorrespondenz verfassen, z. B. E-Mails. B1	29/ It's your turn

13 Tap repair and installation

This unit is about:
- how to replace a washer
- how to replace a cartridge
- how to install a tap

Competences:
- putting illustrations in the correct order
- labelling a diagram
- listening for detail

1 Replacing a washer

The conventional pillar tap can still be found in many homes – particularly in older houses and buildings. It doesn't look as stylish as modern types of taps and mixers, but it has one big advantage – it is the easiest type of tap to repair! Read these step-by-step instructions for replacing a washer carefully.

Before you start: turn off the mains water supply.

1 Remove the cover of the tap. Use an adjustable spanner or blunt jaw pliers for this and pad the jaws with a cloth so that they don't damage the chromium plating.
2 Now you can use a slim spanner and unscrew the headgear.
3 Underneath the headgear is the jumper with the washer.
4 The jumper can be pulled out. If it is stiff, you may need to lever it out with a screwdriver or a coin.
5 Hold the jumper firmly with a pair of pliers and use a small spanner to unscrew the washer-retaining nut.
6 Remove the old washer and clean the jumper.
7 Fit a new washer onto the jumper and replace the washer-retaining nut.
8 Replace the jumper and refit the headgear.

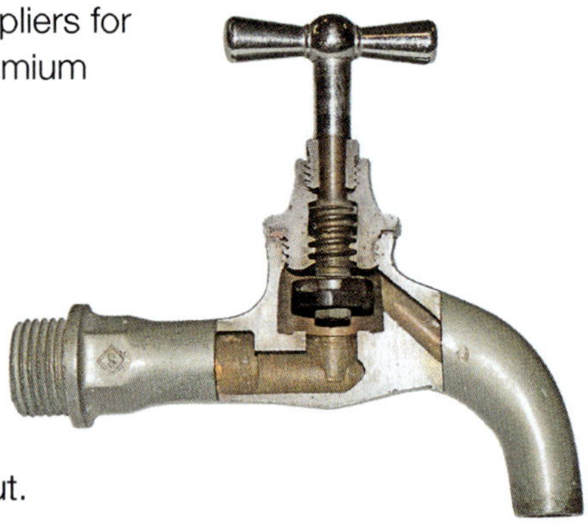
Jim Austin

★ a) The following illustrations are in the wrong order. Match the illustrations to the steps described above and label the parts with the words in blue.

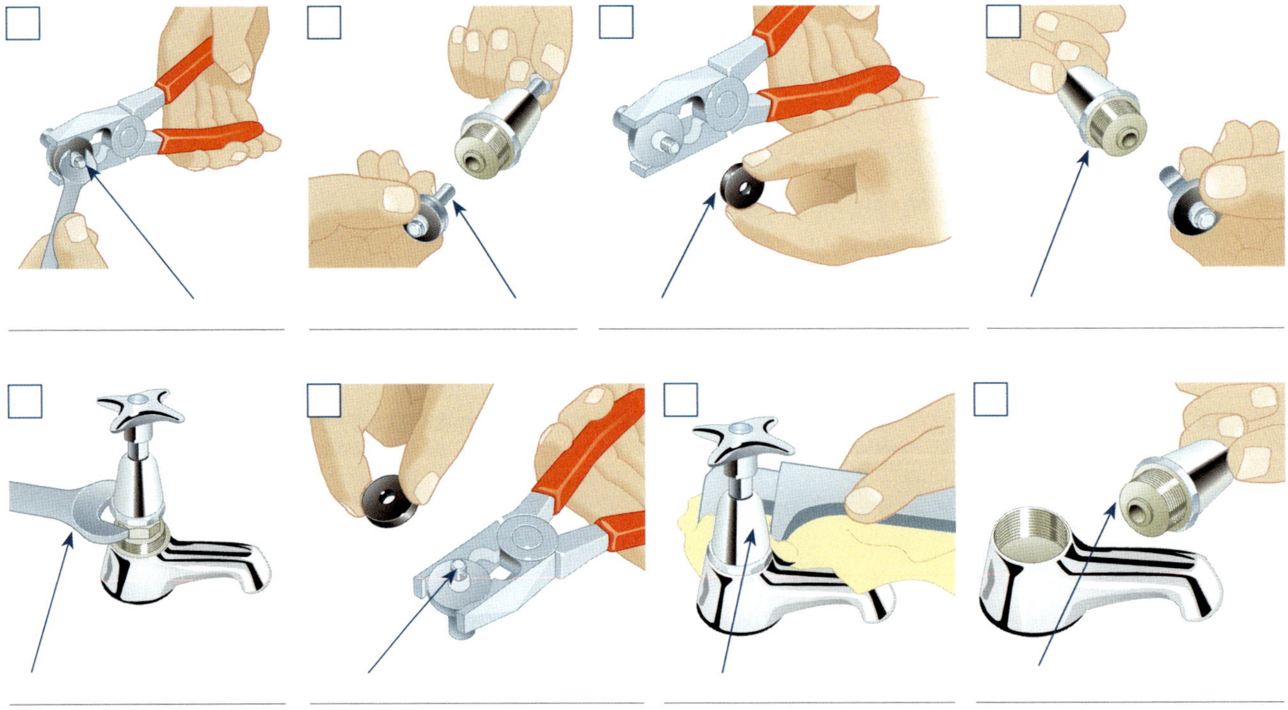

★★ b) Label the illustrations with the words in blue from the instructions.

2 Replacing a cartridge

A single-control tap has a lever that controls water flow and temperature. It is used for kitchen sinks, baths, showers and washbasins. Read the following step-by-step instructions for replacing a cartridge and then label the diagram with the words in blue.

Before you start:
Turn off the hot water system from the cistern or boiler and the cold water supply at the mains.

- Lever out the plug.
- Remove the setscrew with a 3mm Allen key and pull off the lever.
- Unscrew the cap.
- Remove the two retaining screws and remove the complete cartridge.
- Change either the complete cartridge or just the seals.
- Assemble in reverse order. (See diagram.)
- Make sure that the cartridge seals engage in the grooves on the housing.
- Fit the retaining screws and tighten them evenly and alternately.

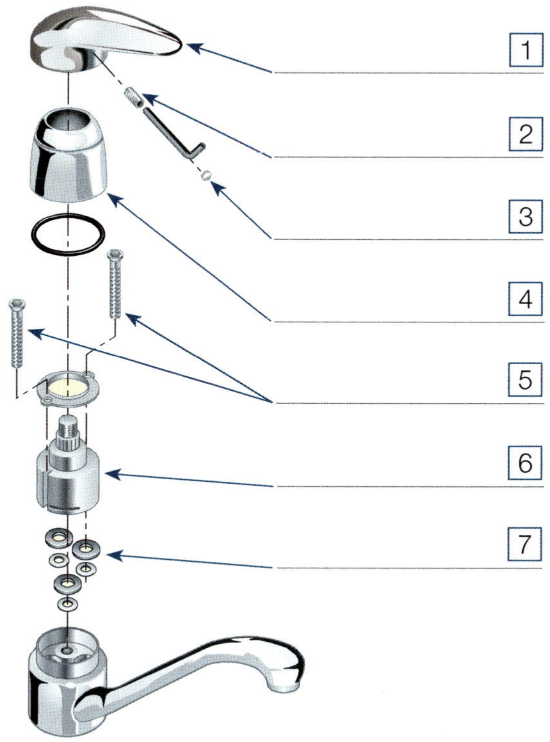

15

3 Installing a tap

A plumber is explaining to you how to install a tap. Listen and then complete the text with words from the box.

clean • flexible • gasket • hole • leaks • nut • spanner • supply • valves

Measure and mark the ______ 1 for the tap. Drill the hole. Fit the two ______ 2 supply lines onto the tap. First make sure that the mounting surfaces are ______ 3. Insert the flexible ______ 4 lines down through the hole. If the tap is fitted with a rubber ______ 5, place it on the bottom of the tap. Set the tap in position and press it down onto the sink. Assemble the washer and counter ______ 6 to fix the tap. Attach the flexible supply lines to the shut-off ______ 7 (left – hot, right – cold) under the sink. Tighten the coupling nuts with a ______ 8. Then turn on the water and check for ______ 9.

IT'S YOUR TURN

Make a list of all the useful vocabulary from this unit. Write the words in a list, write the German word next to them and find pictures to illustrate your words.

14 Taps and mixer taps

This unit is about:
- different types of taps
- a thermostat mixer
- the difference between Fahrenheit and Celsius

Competences:
- listening for information about taps
- writing a short text about a thermostat mixer
- listening for detail

16

1 Different types of taps

There are several types of taps.

a) Listen to a plumber describing the different types to his apprentices and complete the notes below.

Type of tap	Characteristics
conventional pillar tap	old fashioned, capstan handle with …
bib tap	
modern pillar tap	
bath mixer	
sink mixer	
supatap	

★ b) Identify these six taps and label them with the terms above.

A
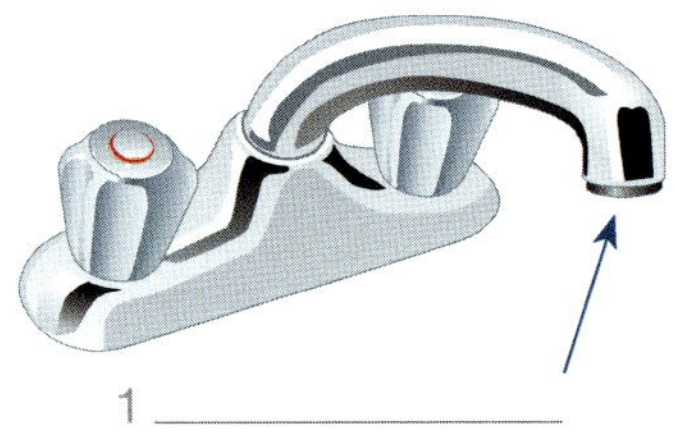

B
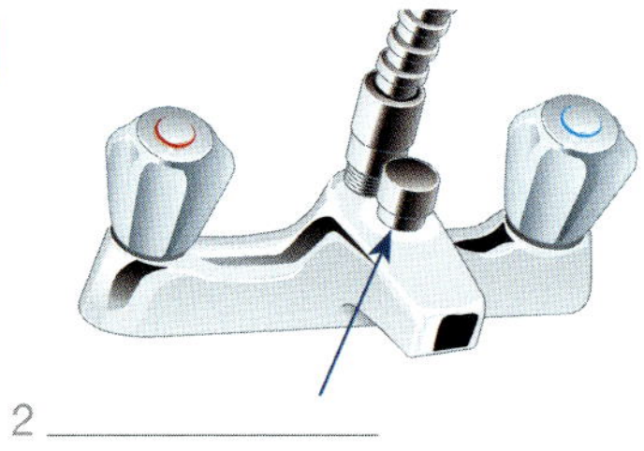

C
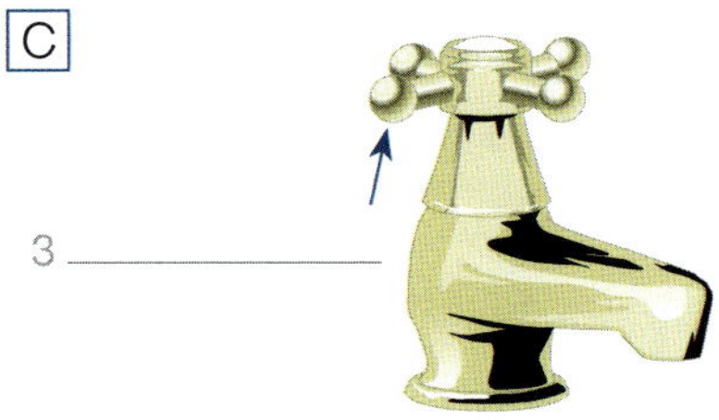

D
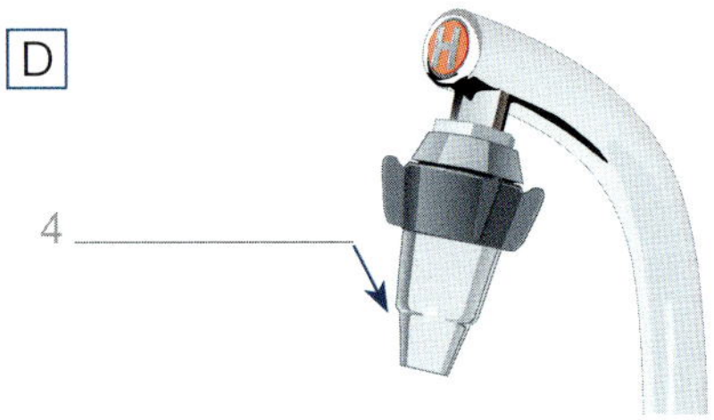

E
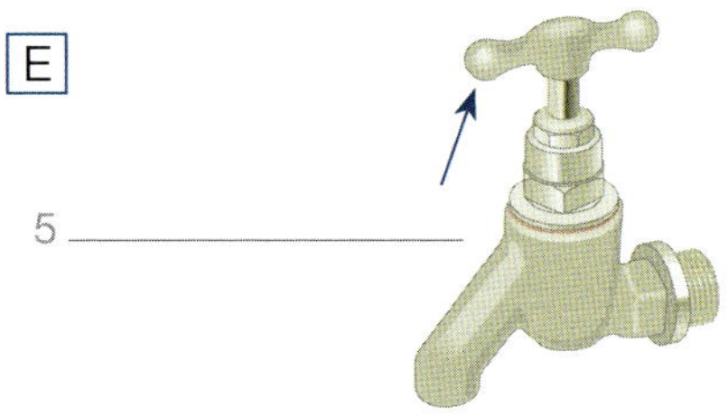

F
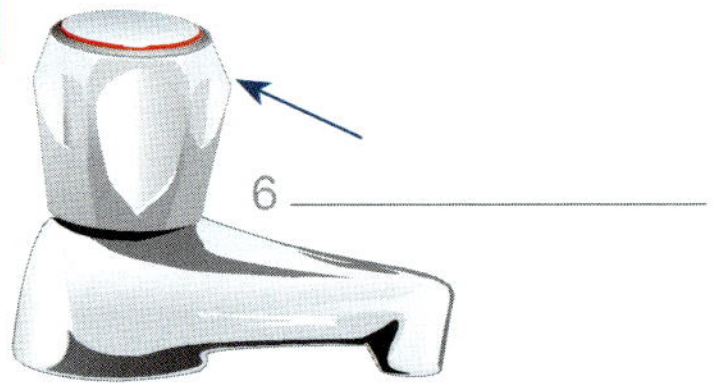

★★ c) Where do the terms in the box go? Label the numbers 1–6 in the illustrations.

button • capstan handle • crutch head handle • handwheel • nozzle • spout

★★★ d) Now write sentences about the taps using your notes and the illustrations.

2 A thermostat mixer

Robert Nadolny, Berlin

A conventional bath or shower mixer has two handwheels or knobs. One is for the hot water and the other is for the cold water. You must adjust both taps until the water from the spout is the right temperature and is also flowing fast enough.
A thermostat mixer looks very similar, but it works differently. With one handwheel you can choose the temperature and with the other you can choose how fast the mixed water flows. It doesn't matter how fast or how slow the water flows, it will always have the exact temperature that you have chosen.

Work with a partner. Discuss what you think the advantages and disadvantages of the conventional mixer and the thermostat mixer are. Take notes and then write down complete sentences. (Use your exercise books.)

Here are some useful words and phrases:

difficult to use	easy to use
exact	inexact
either … or	
too slow	too fast
too hot	too cold
price is expensive/cheap	
service/repair	
simple	complicated

17 🎧

3 How to use a thermostat mixer correctly

A plumber is explaining a thermostat mixer to his apprentice. Listen to the dialogue and complete the sentences with words from the box.

dangerous • eighty • five bar • right • safety stop • thirty-eight • twenty litres • water pressure

1 From the front, the temperature handwheel is on the ____________ of the mixer.
2 The maximum temperature is ____________ degrees Celsius.
3 This can be ____________ for young children.
4 The red button on the temperature wheel is a ____________.
5 It limits the maximum temperature to ____________ degrees Celsius.
6 The handwheel on the left regulates the ____________.
7 The highest pressure that can be set is ____________.
8 The flow at three bar is about ____________ of water a minute.

4 Fahrenheit and Celsius

Some customers only think and talk in Fahrenheit and not in Celsius. Read the text and then convert the temperatures below from Fahrenheit into Celsius.

- Subtract 32 from the Fahrenheit temperature.
- Multiply this number by 5.
- Then divide this number by 9.

1. 120 °F = ____________ °C
2. 98 °F = ____________ °C
3. 50 °F = ______ °C
4. 32 °F = ______ °C

IT'S YOUR TURN

Find out which taps your company usually installs or repairs. How often must the different kinds of taps be repaired? Which tap is the most reliable, which is the least reliable? Write a short text explaining your findings.

15 A private water supply

This unit is about:
- a water pump
- a centrifugal pump

Competences:
- understanding a text about a water pump
- completing a text about a centrifugal pump
- mediating how a pump works

1 A household water pump

Not all houses are connected to the public water system, many have their own private water supply instead.
According to a recent report, around 15 million homes in the USA take their drinking water from private wells or springs, which means they have an individual home well which is partly used for drinking water.

★ a) **Read the text and label the picture below with the words in blue.**

This pump is ideal for households that have a private water supply. The pump is driven by an electric motor. Next to the electric motor is the centrifugal pump. The pump is connected at the front by an inlet pipe with the water source (i.e. a garden well). The pump's outlet pipe is next to the pressure gauge on the top of the pump. It shows the pressure inside the pump. The gauge is connected to a pressure switch. When water is drained, the pressure inside the pump falls. This activates a sensor that switches on the pump. When the operating pressure is reached, the sensor then switches the pump off.

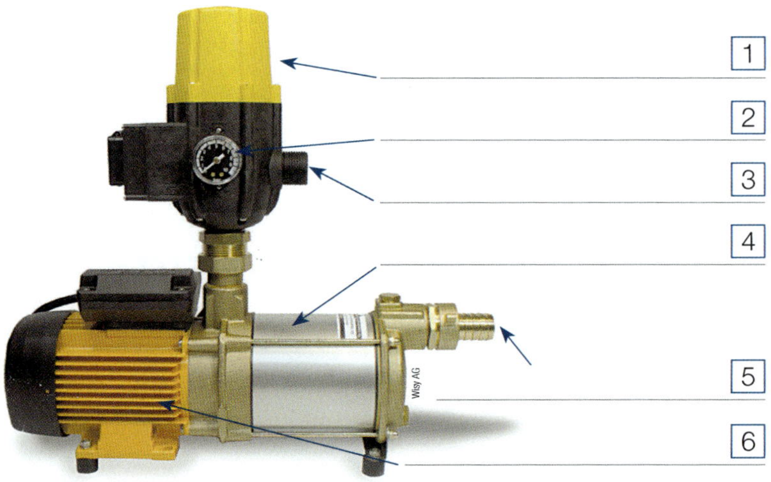

Recently, an environmentally friendly type of this pump has been developed which only needs 0.2W when it is in standby mode.

★★ b) **Read the text again and answer the following questions in complete sentences.**

1 What is mounted next to the centrifugal pump?

____

2 What is on the top of the pump and what is its function?

____

3 What drives the pump?

____

4 What happens when the pressure inside the vessel falls?

____

5 When does the sensor switch off the pump?

____

6 Which parts of the system are connected by the heavy-duty hose?

____

2 A centrifugal pump

As a plumber, you should understand how pumps work.

★ a) Use the key to complete this technical description of the pump. First write down the words with the German translations.

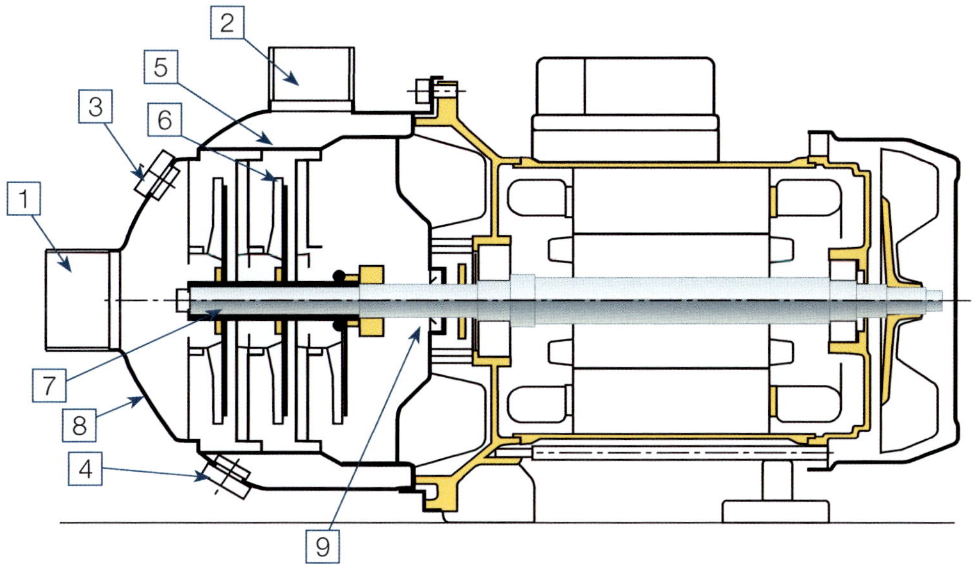

Key

1 horizontal inlet
2 pressure gland
3 vent screw
4 drain screw
5 stage chamber
6 impellers
7 motor-pump shaft
8 pressure casing
9 axial face seal

This pump is a multistage (two-stage) normal suction, horizontal, high-pressure centrifugal pump with a block design with a horizontal ____________[1] and a vertical ____________[2]. The hydraulic section is equipped with several stage ____________[3] (the number varies according to type). There are always the same number of ____________[4]. All the stage chambers and impellers are fitted on a ____________[5]. The pressure ____________[6] surrounding the hydraulic section guarantees a fail-safe seal. All parts that come into contact with water, such as the stage chambers, the impellers and the pressure casing, are made of chromium-nickel steel. The shaft hole through the pump casing is sealed with an axial ____________[7]. The pump must never be allowed to run dry, as this will destroy the seal. Always check that the pump is filled with water. Water can be added by unscrewing the vent ____________[8] and carefully pouring in water. The pump and all the pipes and hoses must be drained of water in the cold season. Exposure to frost can damage the pump. To prevent corrosion the pump should also be drained of water if it is out of service for a long period. The pump is drained by removing the ____________[9] on the underside of the pump.

★★ b) In German, explain to a colleague, who sees your completed text and who does not understand English very well, how a centrifugal pump works. Do not translate every word.

IT'S YOUR TURN

How common are household water pumps which use centrifugal pumps where you live? Research the use of them in your area and write a short blog about it.

16 Waste water disposal

This unit is about:	Competences:
• plumbing blockages • traps • drainage systems	• completing a telephone memo • writing about taps • describing different drainage systems

1 Baths, basins and sinks

18

Baths, basins and sinks together with toilets are the main sources of waste water production in a household. They are also the main sources of plumbing problems because this is where blockages often occur.
Listen to a customer calling Jackson Plumbing and write a telephone memo.

Jackson Plumbing Services	TELEPHONE MEMO
Name of caller:	Date: 14 March 20..
Address:	
Problem:	
Action:	

2 Traps

How much do you know about traps?

★ a) Complete the text with words from the box.

bend • blockages • drains • plastic • sink • smell • water

Every bath, toilet, basin or ______________[1] in a house or flat is fitted with a trap. A trap is simply a ______________[2] in the outlet pipe. The bend fills with ______________[3] and this stops gases from the ______________[4] entering the house and causing a bad ______________[5]. The trap always has some form of access, so that ______________[6] can be cleared. Most modern traps are made of ______________[7] tubing. There are two types of outlet: a 'P' outlet and an 'S' outlet.

★★ b) Which trap has a 'P' outlet and which has an 'S' outlet?

1

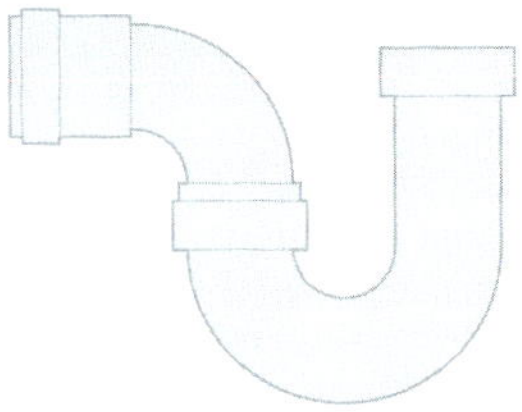

____

2

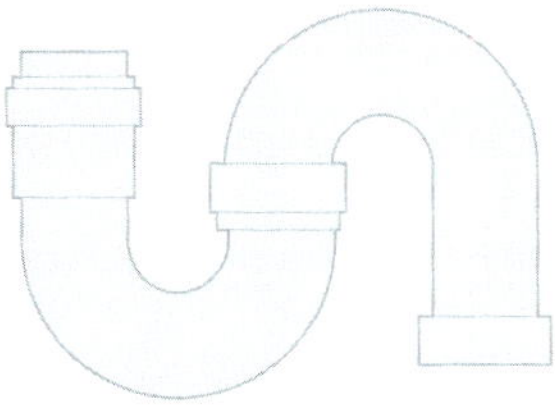

____

3 House drainage systems

Here two drainage systems are described.

★ a) **Read both texts and then label the two illustrations with the words in blue.**

Two-pipe drainage system 1

Many older houses have this system. There's a vertical pipe on an outside wall which carries waste from toilets into the sewer. This is the soil stack. The top of this pipe rises above the eaves of the roof and is open to the air so that sewer gases can escape. This vent is usually protected from birds with a metal guard. Another outside pipe collects the used water from baths, washbasins and showers. This is the waste pipe. There is often an open hopper head at the top of the waste pipe where waste water from various sources is collected. The waste water runs down the pipe to a gully and then into the drains below the house.

1

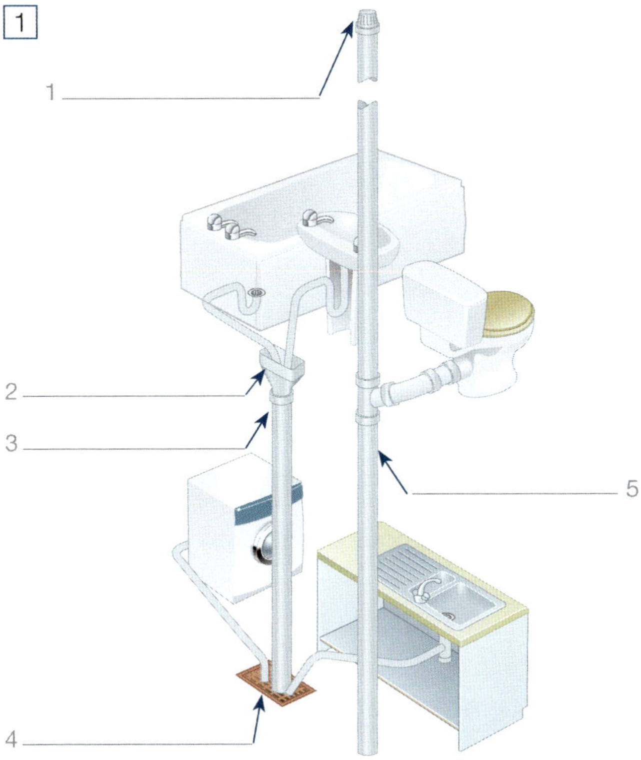

2

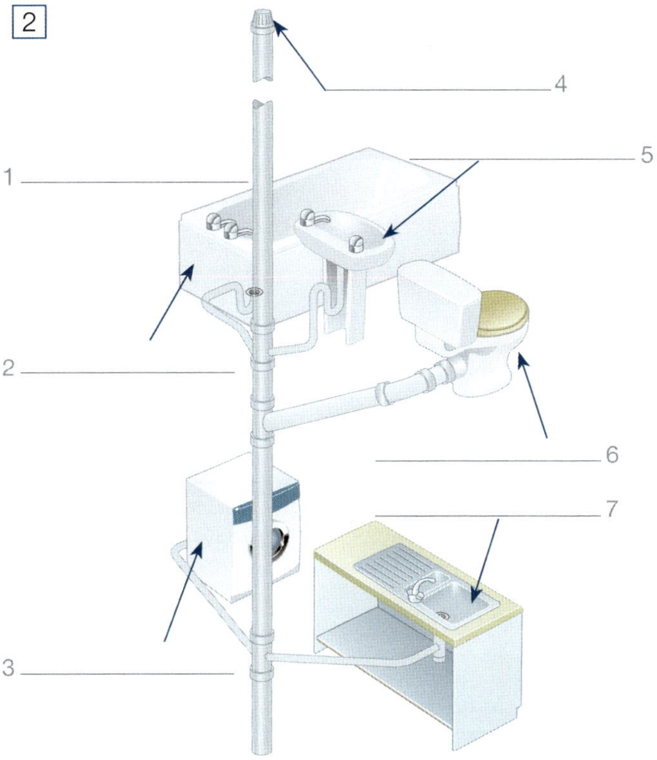

Single-stack drainage system 2

Modern houses have a single-stack drainage system. Waste from all sources – baths, washbasins, showers, toilets, kitchen sinks and washing machines – goes to a vertical pipe. This is the soil and waste stack and it is built inside the house. The waste flows into the sewer and the sewer gases are vented above the roof.

★★ b) **Read the sentences below and say which draining system it is: a two-pipe or single-stack drainage system or both.**

1 There's a vertical pipe on an outside wall. ________________

2 There's a vertical pipe on an inside wall. ________________

3 This draining system has a soil stack. ________________

4 Sewer gases can escape through a vent. ________________

5 The waste water runs down the pipe to a gully. ________________

6 The waste flows into the sewer. ________________

★★★ c) **Work with a partner. What do you think are the advantages and disadvantages of both systems? Write two lists in your exercise books.**

IT'S YOUR TURN

You are not sure when to use a 'P' outlet and when you use an 'S' outlet. Ask your supervisor when you use which outlet. With a partner, prepare and give a short presentation to the class.

Extra practice: Units 13–16

19

1 Replacing a washer

Replacing a washer is a typical task for a plumber. Try and complete the dialogue about how to replace a washer. Then listen to the recording, check your results and fill in any gaps.

Adam: I've got to repair this conventional pillar tap and I have no idea how to remove the ______________[1] of the tap.

John: It's really easy. Look, use this ______________[2] spanner. Be careful and make sure you don't ______________[3] the chromium plating. Good.

Adam: Now what?

John: Unscrew the ______________[4] and you will see the jumper with the little round ______________[5]. Pull out the jumper.

Adam: Oh, it's not easy. It's a bit ______________[6].

John: No problem. Use this fifty-pence ______________[7] to lever it out. Once you have got the washer out, clean the ______________[8] and fit a new washer. Bingo, all done!

Adam: Now I just have to put everything back together again. Thanks.

2 A telephone call

★ a) **Carry out the role-play below with a partner. One of you is a customer and phones Jackson Plumbing.**

Jackson Plumbing	Customer
Melden Sie sich am Telefon mit Ihrem Namen und dem der Firma. Fragen Sie, wie Sie behilflich sein können. →	Sagen Sie Ihren Namen und dass Sie einen Installateur brauchen, da Sie ein Problem im Bad haben.
Fragen Sie nach der Adresse des Kunden / der Kundin. →	Nennen Sie Ihre Adresse.
Fragen Sie nach dem genauen Problem im Bad. →	Sagen Sie, dass das Wasser im … nicht abläuft.
Fragen Sie, ob das schon einmal passiert ist und wie es gelöst wurde. →	Bestätigen Sie, dass dies das … (2., 3,. 4. …) Mal ist. Erklären Sie, wie das Problem beim letzten Mal behoben wurde.
Sagen Sie, dass Sie übermorgen kommen können. →	Fragen Sie, ob es früher möglich wäre.
Vereinbaren Sie einen früheren Termin (Tag und Zeit). →	Bedanken und verabschieden Sie sich.

★★ b) **Unfortunately you cannot deal with the customer so you write a telephone memo for your colleague.**

Jackson Plumbing Services	TELEPHONE MEMO
Name of caller:	Date: 22 March 20..
Address:	
Problem:	
Action:	

▶▶▶ ICH KANN

	Seite/Übung
☐ Ich kann berufs- oder fachbezogene Sachverhalte und Prozesse (z. B. *outlets*) einfach, aber strukturiert beschreiben, erklären bzw. zusammenfassen. B1	41/ It's your turn
☐ Ich kann ein Gespräch über mir vertraute Themen beginnen, in Gang halten und beenden. B1	37/2
☐ Ich kann kurzen und deutlich gesprochenen Texten und Tonträgern die wesentlichen Informationen entnehmen, wenn es um vorhersehbare alltägliche oder berufsspezifisch vertraute Dinge geht. A2	35/3
☐ Ich kann Interviews, Berichten und Vorträgen in den wesentlichen Punkten folgen. B1	36/1a
☐ Ich verstehe umfassende Anweisungen, Fragen und Auskünfte in Alltags- und Berufssituationen sowie auf Reisen. B1	40/1
☐ Ich verstehe Arbeitsanleitungen in meiner Ausbildung und am Arbeitsplatz sowie schriftliche Anmerkungen zu meinen eigenen Arbeiten. A2	34/1
☐ Ich kann Sachtexte verstehen, die einfach und für ein breites Publikum geschrieben sind und in denen es um Themen geht, die mit meinen Interessen oder meinem Fachgebiet in Zusammenhang stehen. B1	39/2
☐ Ich kann Informationen in verschiedenen längeren Texten oder Textteilen finden, die ich brauche, um eine bestimmte Aufgabe zu lösen. B1	38/1
☐ Ich kann mit Hilfe von vorgegebenen Wörtern, Illustrationen und Filmmaterial eine kurze, einfache Geschichte schreiben oder Vorgänge und Abläufe beschreiben. A2	36/1d
☐ Ich kann einfache beschreibende Texte zu grundlegenden Sachverhalten oder Aufgabenstellungen aus meinem Fachgebiet verfassen. B1	37/ It's your turn
☐ Ich kann einfache private Korrespondenz sowie berufliche Standardkorrespondenz verfassen, z. B. E-Mails. B1	39/It's your turn; 40/1

17 Shower systems

This unit is about:
- gravity-fed showers
- electric showers

Competences:
- listening and note taking
- labelling diagrams of different shower systems
- writing a blog

20 1 Choosing a shower system

Mr and Mrs Bates want to install a shower in their bathroom. Mr Pearson, a local plumber, is giving them some advice.

★ a) Listen to the dialogue and take notes.

Name of shower system 1: ______________________________

Details ______________________________

____

____

____

____

Name of shower system 2: ______________________________

Details of type A ______________________________

____

____

____

____

Details of type B ______________________________

____

____

★★ b) Use your notes to answer these two questions.

1 Which shower system do you think will be the most expensive to install and why?

____

____

____

2 Which shower system do you think will be the cheapest to install and why?

____

____

____

★★★ c) Listen again and tick the words or expressions that you heard and explain them.

☐ a pretty bathroom	☐ brochure with an illustration	☐ in the loft
☐ cheap job	☐ storage tank	☐ near to the mains water pipe
☐ that means nothing to me	☐ instant hot water	☐ lots of time to install

2 Three shower systems

★ a) Label the diagrams with the words from the box. The recording and the notes you made in exercise 1a will help you.

booster bump • cold-water cistern • gravity-fed shower • hot-water storage cylinder • immersion heater • shower hose • wall unit • instant electric shower, type A • instant electric shower, type B • waste pipe

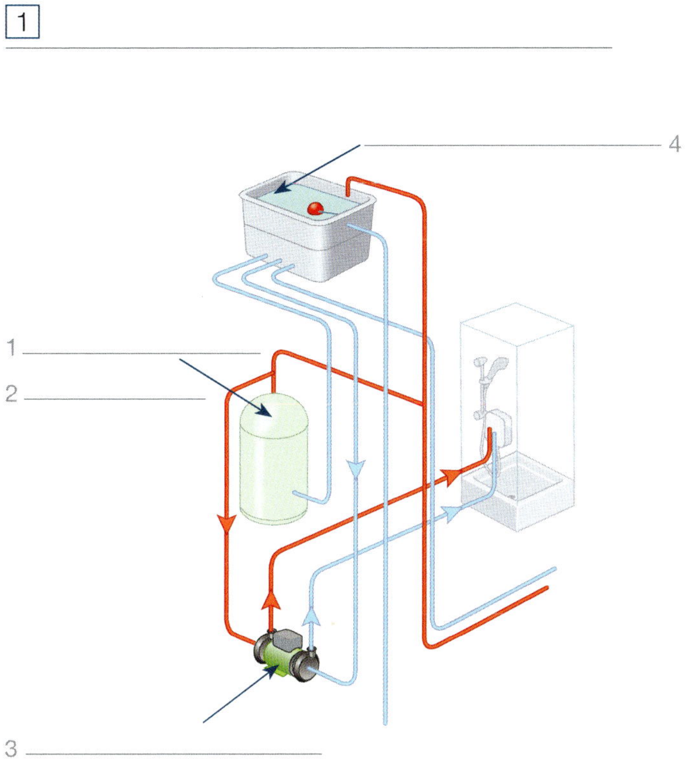

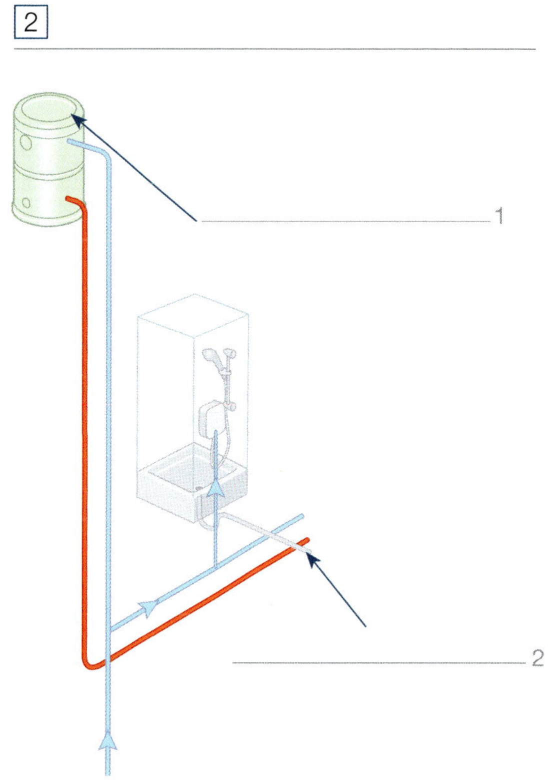

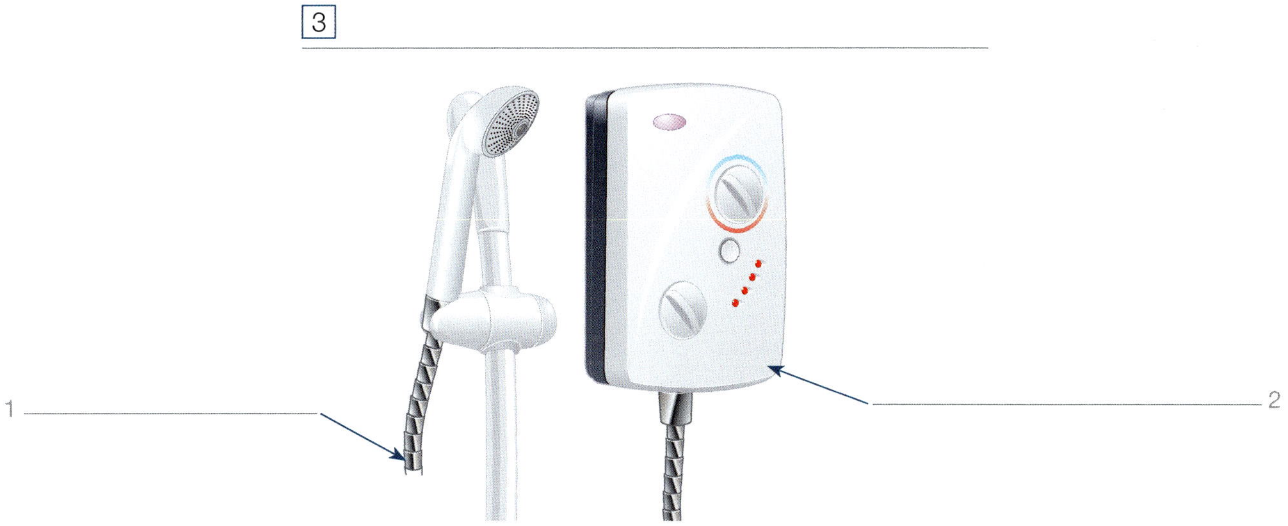

★★ b) Work with a partner. Student A describes one of the showers and student B has to say which shower type it is. Then swap roles.

IT'S YOUR TURN

Your boss has asked you to find out about the different types of shower in your country. Do the research and find out which is the most popular and write a short blog about it.

18 Heating water

This unit is about:
- water-heating systems
- features of water heaters

Competences:
- labelling a diagram about water heaters
- understanding texts about water heaters
- understanding definitions

1 Water-heating systems

There are three main water-heating systems.

★ a) Look at the diagrams of the three systems for providing warm water. Decide which type of system they are and label them.

A centralized system consists of a water heater which delivers warm water to appliances around the house. There are two kinds of decentralized systems: single-point (where every appliance has its own water heater) and multi-point (where a small group of appliances share a water heater).

1 ______________________

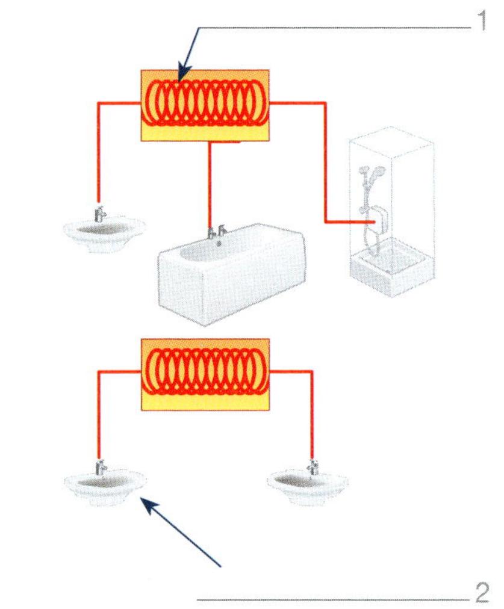

2 ______________________

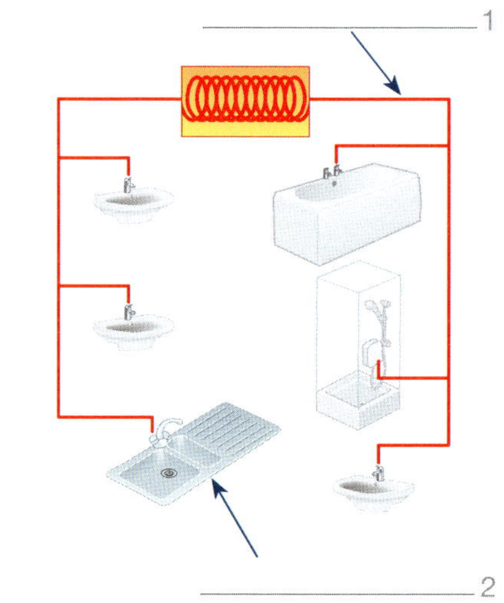

3 ______________________

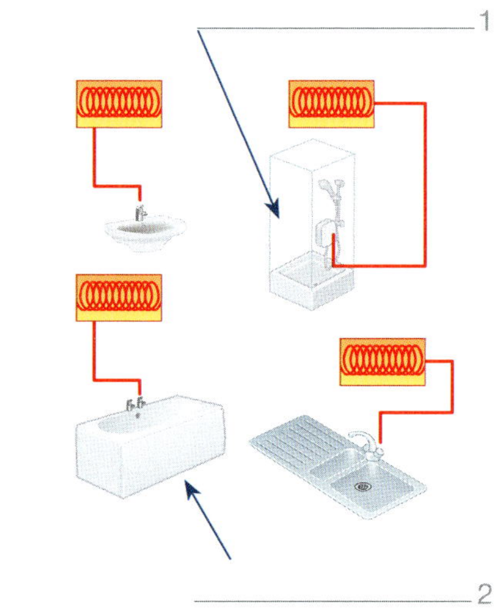

★★ b) Now label these items in the diagrams.

bathtub • heater • kitchen sink • pipework • shower • washbasin

★★★ c) Read the list of system features. Are they about centralized (C) or decentralized (D) water-heating systems?

1 Because of low water pressure and heat loss, these systems need circulation pumps and must be very well insulated. ☐

2 The cost of buying and maintaining several heating units for a house or a building can be high. ☐

3 Heat loss is minimal because the heaters are close to the taps. ☐

4 Can be complicated to install. ☐

5 Hot water can take some time to reach the taps. ☐

6 Easy to install. ☐

7 There will be some heat loss because of the distance the water has to travel. ☐

2 Water heaters

★ a) These two texts are about two different types of water heaters. Write the type of water heater above the texts.

Continuous flow heater • Storage heater

This heater does not heat water in a tank. Water is taken from the building's water supply and is heated as it passes through the heater.	Cold water from the mains is heated in an insulated tank and it stays there until it is used. The hot water that has been used is replaced by cold water, which is then heated.

★★ b) Think about how the two water heaters work and then look at the operating features in this table. Underline the features you think are correct for each type of water heater.

Operating features	Storage heater	Continuous flow heater
Necessary heat flow	low / high	low / high
Heating time	depends on heater output / immediate	depends on heater output / immediate
Hot water supply quantity	limited depending on storage volume / unlimited	limited depending on storage volume / unlimited
Filling time in minutes for 1 full bath	5–7 / 15–20	5–7 / 15–20
Heat loss Without insulation	very high / limited / practically none	very high / limited / practically none
With insulation	very high / limited / practically none	very high / limited / practically none

3 Definitions

It is useful to be able to explain basic plumber words to a customer. Find words from this unit that fit these definitions.

1 A system that isn't controlled from a central point: __________

2 The movement of water around a closed system: __________

3 A machine that moves water around a system: __________

4 Covered with a material that slows down heat loss: __________

5 A water outlet that can be turned on or off by hand: __________

6 The large pipe that delivers water to a house: __________

IT'S YOUR TURN

Work with a partner. Choose one of the situations below and take turns. A customer comes into your shop and wants to know

- the difference between storage heaters and continuous flow heaters.
- if it would be better to install a single-point water heating system or a multi-point system.

Think about what the customer has to consider before making a choice.

19 Working with piping

This unit is about:
- pipe fittings
- bending pipes
- pipe insulation

Competences:
- listening to a record and labelling diagrams
- completing a gap text
- talking about insulation materials

21

1 Pipe fittings

A plumber is describing some pipe fittings to his new apprentice. Listen and label the diagrams with the words from the box.

equal-size connector • 90° elbow • gate valve • slow bend • T-junction

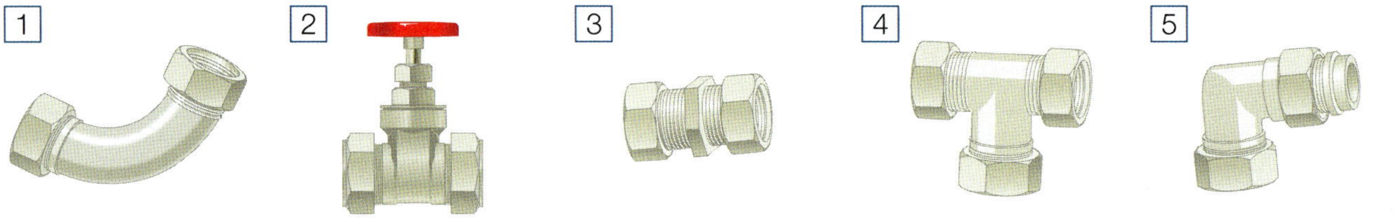

1 ______ 2 ______ 3 ______ 4 ______ 5 ______

2 Bending pipes

★ a) Which tools are used to bend pipes? Read these descriptions and match them with the pieces of equipment in the illustrations in exercise 2b.

1 This is a handy piece of equipment that plumbers always carry with them. It is suitable for bending low gauge soft piping made, for example, from copper. It needs two hands to operate it. ☐

2 This tool is portable and it can be taken on jobs. It is used to bend piping made of hard metals such as steel. ☐

3 This is a heavier piece of equipment which stays in the plumber's workshop and is used to bend piping made of hard metals. ☐

4 This is a handy power tool and it can be taken on jobs. It allows a plumber to bend pipes easily, but without using his/her own strength. ☐

★★ b) Now label the pipe-bending equipment with the words from the box.

double-handed pipe bender • electric pipe bender • pipe-bending machine • portable hydraulic pipe bender

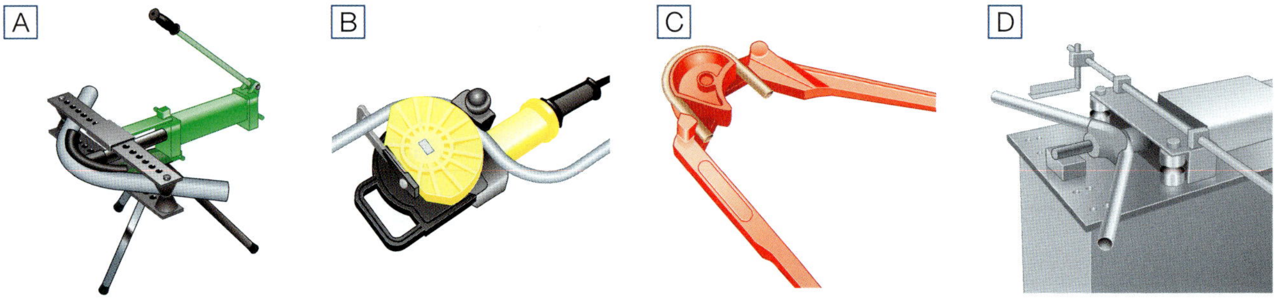

A ______ B ______ C ______ D ______

3 Insulating pipes

Say which of these statements about insulating pipes are true (T) or false (F). Then compare your answers with a partner.

We insulate pipes in order to …

		T	F
1	… stop them from freezing and splitting in very cold weather.		
2	… hold them together.		
3	… stop cold water warming up if it is near sources of heat, such as hot-water pipes.		
4	… reduce the noise of water or air running through the pipes.		
5	… stop water leaking out.		
6	… stop the pipes from overheating in summer.		
7	… stop the build-up of moisture on cold-water pipes in warm weather.		
8	… stop hot water cooling down as it runs through the pipes.		

4 Pipe insulation materials (lagging)

There are different kinds of insulation materials.

bends • fix • flexible • metallic • rolls • wrap

★ a) Use words from the box to complete the text.

Self-adhesive foam wrap is 50 mm thick and comes in ________________ [1] usually five or ten metres long. Some types have a ________________ [2] finish. Before you ________________ [3] the lagging into place, make sure that the pipes are dry and clean.

Peel the backing paper off and ________________ [4] the material around the pipes. Overlap the foam wrap as you go, especially at ________________ [5]. This ________________ [6] lagging is also useful for insulating fittings with difficult shapes, such as stopcocks.

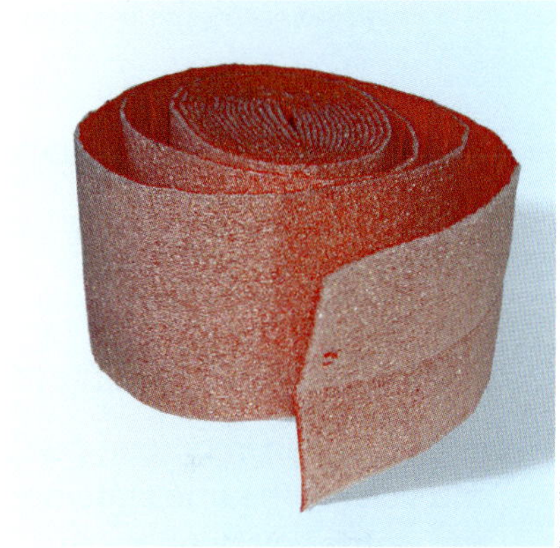

Vanessa Maly

★★ b) The sentences for these instructions are in two parts. Join the correct parts together.

1 Plastic foam tubes are available
2 The tubes are split down one side and have to be
3 They are secured at intervals
4 If the building is in an area which often has severe frosts,
5 The thickest possible lagging should also be used if

A it is worth using the thickest lagging possible.
B eased open to fit around the pipe.
C the piping is particularly exposed.
D in several thicknesses.
E with adhesive tape or with special clips.

Fotolia.com/Nomad_Soul

1		2		3		4		5	

IT'S YOUR TURN

You want to tell the class about the different kinds of pipe insulation material used at your place of work. Write a list and explain what each type of insulation is used for. Report back to the class.

20 Central heating

This unit is about:
- central heating pumps
- changing a pump

Competences:
- understanding statements about pumps
- completing instructions
- producing a mind map

1 Air and water

A central heating system can use either warm air or warm water. Warm water systems are by far the most common type. They use four main devices for regulation: a thermometer, a manometer, a temperature regulator, and a pressure regulator.
Which devices do these definitions describe?

1 This device prevents the system heating up or cooling down too much.
This is a ______________________________ .

2 This device prevents pressure in the system building up to dangerously high levels.
______________________________ .

3 This device measures how warm the water in the system is.
______________________________ .

4 This device measures the pressure in the system.
______________________________ .

Vanessa Maly

2 Central heating pump

Pumps are an important part of all central heating systems. Research this and tick (✓) the true statements. Use a dictionary to find the meanings of any words you don't know.

		T	F
1a	The pump should only contain clean water which is free of air bubbles.		
1b	The pump will not be damaged by water which is dirty and contains air bubbles.		
2a	Wet central heating pumps should be positioned horizontally.		
2b	Wet central heating pumps should be positioned vertically.		
3a	Dry central heating pumps should never be positioned with the motor underneath.		
3b	Dry central heating pumps should always be positioned with the motor underneath.		
4a	It is unnecessary to protect the pump from drips of water and condensation.		
4b	The pump should always be protected from drips of water and condensation.		
5a	The pump should never be run when it is dry.		
5b	The pump can occasionally be run when it is dry.		

3 Changing a central heating pump

★ a) Here is a step-by-step guide to changing a central heating pump.
Look at the illustrations and complete the text below with the phrases from the box.

- in place and then remove it.
- using the valve handle or an adjustable spanner.
- to catch any water that escapes when you remove it.
- to remove any moisture and then reconnect the electrical system.
- washers in the unions to stop leaks.
- supply to the central heating system controls.

1 Turn off the electricity

2 Close down the service valves on each side of the pump

3 Put a bucket and towels underneath the pump

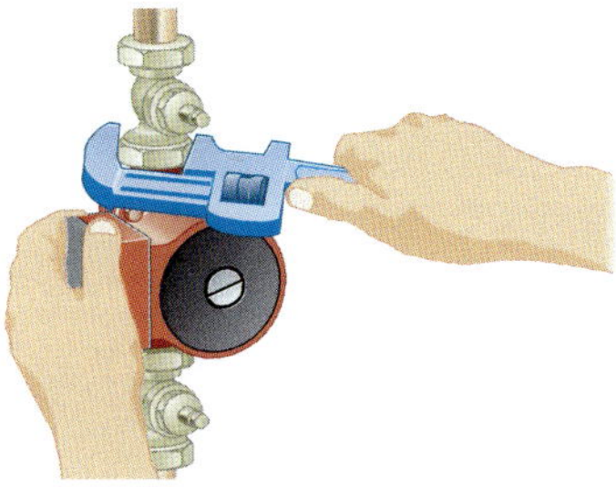

4 Unscrew the union nuts that hold the pump

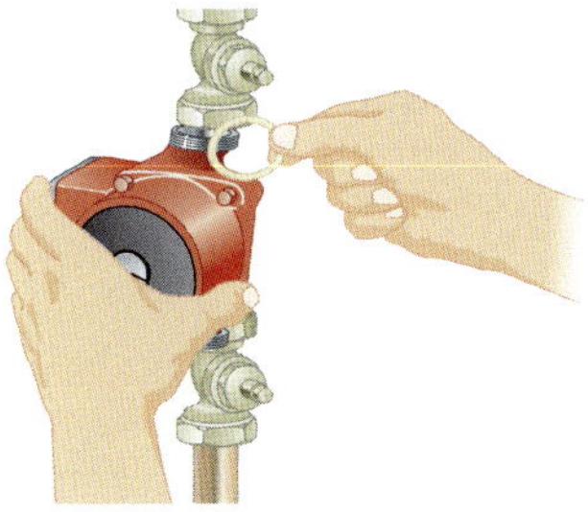

5 Fit the new pump in place with new sealing

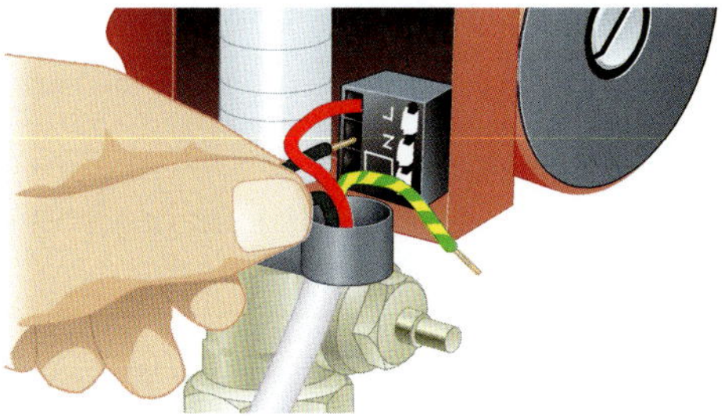

6 Dry the pump carefully

★★ b) Imagine you have changed a central heating pump. Write what you did in your exercise book in the past tense and use linking words, e.g. First I … / After that I … / When I … / Then … / Finally, …

IT'S YOUR TURN

Make a mind map of all the important vocabulary from this unit. Write both the English words and the German translations beside them. Then compare your mind map with a partner's.

Extra practice: Units 17–20

1 Shower systems

You have found a new shower in a catalogue. Now you want to convince your boss that you should start to offer it to your customers.
Work with a partner. Use the help below to work out a dialogue. Present it to the class.

You	Your boss
Sagen Sie Ihrem Chef / Ihrer Chefin, Sie haben eine neue Art Dusche in einem Katalog gesehen.	Fragen Sie, was so neu an der Dusche ist.
Erklären Sie, dass es die Vorteile einer durch Schwerkraft betriebenen Dusche wie auch die eines elektrischen Durchlauferhitzers kombiniert.	Fragen Sie nach Details.
Erklären Sie Einzelheiten: wie das Wasser erhitzt wird, wo sich der Tank befindet, dass man keine Pumpe braucht, dass … .	Fragen Sie, ob Sie sich den Katalog anschauen dürfen.
Erklären Sie dem Chef / der Chefin, dass Sie den Katalog einem Freund geliehen haben, aber ihn in zwei Tagen / nächste Woche / … wiederbekommen.	Fragen Sie, ob es schneller geht.
Sagen Sie, dass das leider nicht geht. Entschuldigen Sie sich.	Erklären Sie, dass Sie natürlich zuerst die Dusche im Katalog sehen müssen, bevor Sie sich dafür entscheiden können.

2 Water-heating systems

Your boss has asked you to help a British colleague. Write down the equivalent English sentences for these German ones.

1 Umwälzpumpen müssen sehr gut isoliert sein.

____

2 Es gibt keine Aufwärmzeit, das warme Wasser kommt sofort.

____

3 Der Wärmeverlust bei isolierten Warmwasserspeichern ist niedrig.

____

4 Die Anschaffungs- und Wartungskosten für ein Warmwassersystem können hoch sein.

____

5 Das Wasser wird im Durchlauferhitzer aufgeheizt.

____

6 Es gibt eine begrenzte Menge an Warmwasser.

____

3 Central heating systems

You want to learn for your exam and your American cousin is staying with you. He helps you by asking you questions. Answer his questions. Try not to look at the units.

1 What two elements can a central heating system use to produce heat? ____________
2 Which system is the most common type? ____________
3 Apart from a manometer, a temperature regulator, and a pressure regulator, what other device does a central heating system use for temperature regulation? ____________
4 What could damage the pump? ____________
5 What should you always protect the pump from? ____________
6 Can you run the pump when it is dry? ____________
7 What must you do first if you want to change a central heating pump? ____________
8 Why should you put a bucket and a towel under the pump when you are changing it?
____

4 Pipe insulation material

Because you understand plumbing, a friend has asked you about the advantages of pipe insulation. Write a blog of about 80 words saying why you should insulate pipes and what you would use.

▸▸▸ ICH KANN

		Seite/Übung
☐	Ich kann Sachverhalte (z. B. Arbeit) und Abläufe mir vertrauter Situationen (z. B. *insulation material*) beschreiben. B1	49/ It's your turn
☐	Ich kann in einfachen zusammenhängenden Sätzen Diagramme beschreiben und interpretieren. B1	45/2b
☐	Ich kann eine Rolle in einem Dialog oder einer Simulation übernehmen (z. B. Berufssituation). A2	47/ It's your turn
☐	Ich kann Anweisungen, Fragen und Auskünfte in Alltags-, Ausbildungs- und Berufssituationen im Wesentlichen verstehen. A2	48/1
☐	Ich verstehe umfassende Anweisungen, Fragen und Auskünfte in Alltags- und Berufssituationen sowie auf Reisen. B1	44/1
☐	Ich kann einfachen Alltagstexten, die sich auf vertraute Situationen beziehen, die wichtigsten Informationen entnehmen (z. B. berufliche Alltagsbeschreibungen). A2	49/3
☐	Ich kann Sachtexte verstehen, die einfach und für ein breites Publikum geschrieben sind und in denen es um Themen geht, die mit meinen Interessen oder meinem Fachgebiet in Zusammenhang stehen. B1	47/2
☐	Ich kann Informationen in verschiedenen längeren Texten oder Textteilen finden, die ich brauche, um eine bestimmte Aufgabe zu lösen. B1	50/2
☐	Ich kann aus vorgegebenen Wörtern einfache Sätze bilden und aufschreiben. A1	49/4b
☐	Ich kann mit Hilfe von vorgegebenen Wörtern, Illustrationen und Filmmaterial eine kurze, einfache Geschichte schreiben oder Vorgänge und Abläufe beschreiben. A2	51/3a
☐	Ich kann einfache beschreibende Texte zu grundlegenden Sachverhalten oder Aufgabenstellungen aus meinem Fachgebiet (z. B. Arten von Duschen) verfassen. B1	45/It's your turn; 51/3b

21 Radiators

This unit is about:

- types of radiators
- parts of a radiator
- replacing a radiator valve

Competences:

- listening for information
- describing different types of radiators
- putting a jumbled step-by-step guide in order

There are four main types of radiators. They differ in the way they work and the materials they are made from. Almost all radiators are still made of metal, steel, cast iron or cast aluminium. These differences affect the appearance, the efficiency (heat output), the durability and the price of the different types of radiators. A customer buying a radiator must consider all these factors and needs expert advice.

22 1 Choosing a radiator

A customer wants to find out about different types of radiators at a plumbing supplies shop.

★ a) **Listen to the dialogue and then match the sentence halves below to make complete sentences which describe the different types of radiators.**

1 Steel radiators are
2 The disadvantage of steel radiators
3 Cast metal radiators
4 One of the main advantages of cast metal radiators
5 Pipework radiators normally look
6 Pipework radiators are
7 Pipework radiators have a higher
8 The amount of heat that panel radiators give off

A heat output than other radiator types.
B not expensive to buy.
C is that they corrode easily.
D is that they don't corrode easily.
E depends on their size.
F more expensive than cast metal or steel radiators.
G are made of cast iron or aluminium.
H more attractive than other types of radiators.

★★ b) **Label the radiators below. The words in the box will help you.**

cast metal • panel • pipework

1

Shutterstock/andrei_sv

____

2

Shutterstock/revers

____

3
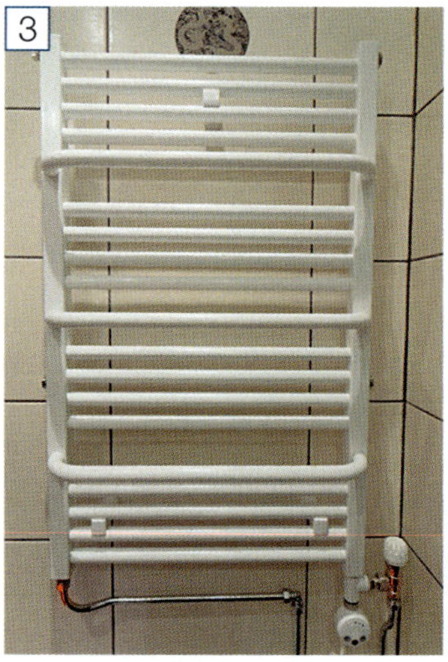
Fotolia.com/Piotr Wawrzyniuk

____

★★★ c) **Now sum up what you know about the three types of radiators in a table.**

2 Parts of a radiator

A customer wants to know what the different parts of a radiator are for.

★ a) Match the descriptions of the parts of a radiator (1–3) to the parts shown on the diagram (A, B, C).

1 This part is used to adjust the flow of water into the radiator. It controls how hot the radiator is.
2 This part is used to balance the system. If one radiator in the system is too hot or too cool, adjusting this knob will bring it back into balance with the rest of the system.
3 This part is used to release airlocks in the radiator. Bleeding radiators to release airlocks will make them heat a room more efficiently.

1		2		3	

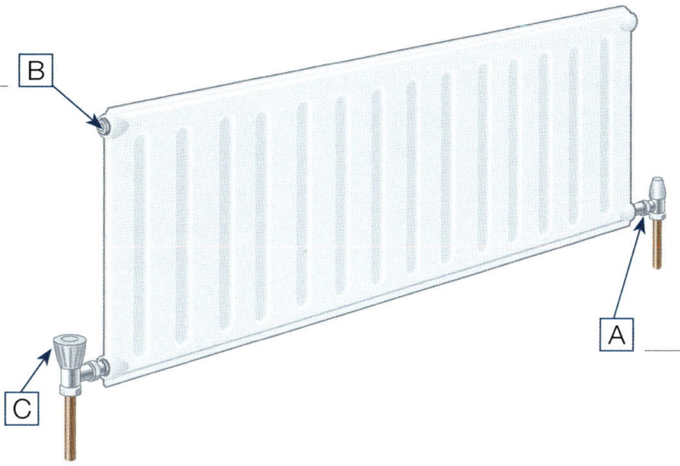

★★ b) Now label the three parts of the radiator with these terms.

bleed valve • lockshield valve • manual handwheel

3 How to replace a radiator valve

This step-by-step guide is mixed up. Put the steps into the correct order.

A Undo the cap nut that connects the pipe to the valve: turn it clockwise. Lift away the valve and allow the cap nut to slip down the pipe. ☐
B Drain the heating system. ☐
C Undo the new valve with a radiator spanner if it doesn't fit the old adaptor. Use wire wool to clean the threads inside the ends of the inlet and outlet pipes. ☐
D Undo the nut that connects the valve to the radiator: turn it anticlockwise. Stop the valve turning by holding the body of the valve upright with a second spanner. ☐
E Then tighten it further with a spanner. Hold the valve body tight with the second spanner as you do this. ☐
F Refill the system and bleed all of the radiators to let out trapped air. ☐
G Thread the cap nut on the adaptor onto the end of the new valve and tighten it with your fingers. ☐
H Slide up the cap nut and connect it to the valve. ☐
I Screw the new adaptor into the radiator. ☐

IT'S YOUR TURN

Work with a partner. You want to renew all the radiators in your apartment. You have chosen one type of radiator but your partner prefers another type. Use the information in this unit to try to convince your partner that your choice is the better one.

22 Solar power

This unit is about:
- solar power
- heating water using solar power

Competences:
- talking about advantages and disadvantages of solar power systems
- identifying problems in solar heating

1 Solar power in buildings

Shutterstock/zstock

Is solar power the answer to the energy crisis we are facing? Here is some information about the use of solar power in private houses.

★ a) **Match the sentence halves in the table below to write about the advantages and disadvantages of using solar power for heating.**

A solar power system …

1	doesn't pollute	A	energy.
2	can't produce any power at night,	B	in winter and in cloudy weather, so a back-up power supply is needed.
3	doesn't need much	C	easily just by adding more solar panels.
4	produces much less power	D	unpleasant smells.
5	costs less now than in the past, and in the future	E	the environment.
6	produces free	F	they will be even cheaper.
7	doesn't need any	G	noise.
8	can be expanded	H	install.
9	doesn't make any	I	so storage batteries are needed.
10	doesn't produce any	J	fuel.
11	is easy to	K	maintenance.

1 A solar power system doesn't pollute ______________________

2 It can't produce any power at night, ______________________

3 It doesn't need much ______________________

4 It ______________________

5 It ______________________

6 It ______________________

7 It ______________________

8 It ______________________

9 It ______________________

10 It ______________________

11 It ______________________

★★ b) **Work with a partner. Discuss the advantages and disadvantages of solar power systems.**

A solar power system is …
Yes, but don't forget that …
True, but what about …?
That's a good point. However, a solar power system …
What about the cost? Isn't a solar power system …?

2 Problems in heating water with a solar collector

The following things can cause problems, so it is important that plumbers are aware of them. Write them in the correct part of the table below.

1. There's a vapour lock in the collector. This can be removed by allowing the system to cool and then draining it.
2. The hot-water taps are switched on and off too often.
3. The weather is cloudy.
4. There's no flow of water through the collector because it's blocked (due to scaling, for example).
5. The surface of the collector is dirty.
6. There's no cold-water supply (due to a blockage, for example).
7. The collector may be completely in the shade so no sunlight can reach it.
8. Too much hot water is being used.
9. There's an airlock in the pipes.

Problem	Cause of problem
No water comes out of the taps. 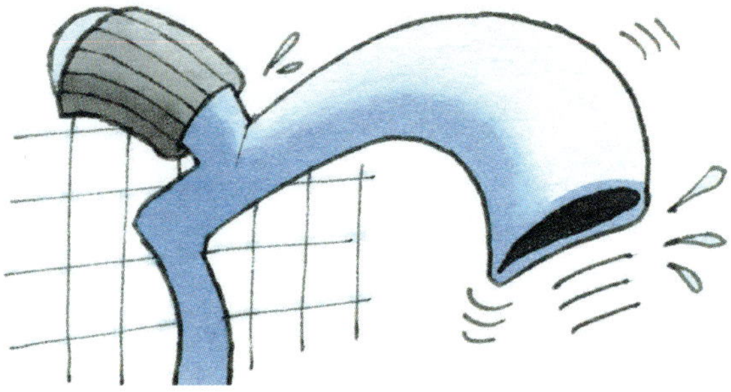	
The water isn't heated at all, although the cold-water flow is normal. 	
The water isn't hot enough. 	

IT'S YOUR TURN

Your neighbour wants to install solar heating in his house and has asked your advice. Research all the factors that are involved in solar heating for private houses. Present your findings to the class. You could do a Power Point presentation.

23 Air circulation, ventilation and air conditioning

This unit is about:
- air circulation
- ventilation
- air conditioning

Competences:
- reading about systems that control air quality
- labelling parts of an air conditioning system
- listening for information

1 Air quality in buildings

Shutterstock/Shi Yali

★ a) **The following text is about systems to control air quality. Find words in the text with the meanings below.**

The same technology that prevents heat loss in modern buildings can also cause poor air quality.
In houses or offices poor air quality can mean that the humidity is too high – there is too much moisture in the air – or that there is dust in the air. This can also be a big problem in industrial buildings. Smoke, toxic materials and fibres in the air can also be a problem here.
That's why systems to control air quality in a building are so important. There are three types of systems which do this. The first of these is the air-circulation system. This does not bring air in or out of a building. Instead it makes air flow around individual rooms or all of the building. The second type is the ventilation system. This allows air to flow in and out of a building. It also filters the air.
Finally, there is the air conditioning system. Air conditioning cleans the air in a building and keeps the temperature and humidity at a constant level. In houses and offices, air conditioning keeps the temperature at a level that is comfortable for people to live and work in.

1 Very small drops of water in the air: ______________________

2 A fine powder of sand, earth and other materials: ______________________

3 The grey or black gas which is produced when something burns: ______________________

4 Something that is poisonous: ______________________

5 The movement of air around a building: ______________________

6 A process which allows fresh air to enter a room or a building: ______________________

★★ b) **Read the text again and say whether the statements are true (T) or false (F). Then correct the false statements.**

		T	F
1	Air moves easily in and out of modern buildings without any help.		
2	High levels of humidity are a problem in buildings where people live and work.		
3	Air-circulation systems bring air from outside into a building.		
4	Ventilation systems keep the humidity in a building at the correct level.		
5	Air conditioning systems are designed to control how warm or cool a building is and how much moisture there is in the air.		

____

____

____

★★★ c) **Sum up the text in no more than 60 words.**

2 An air conditioning system

Look at the diagram of an air conditioning system and read the descriptions of how the parts function. Complete the descriptions with words from the box.

air filter • blower • compressor • condenser • condenser coils • evaporator coils • fan • refrigerant

1 ______________________
These turn the refrigerant into a low-temperature gas.

2 ______________________
This a chemical used by the system to cool the air down.

3 ______________________
This blows air out of the condenser.

4 ______________________
This the part of the air conditioning system which sits outside the building.

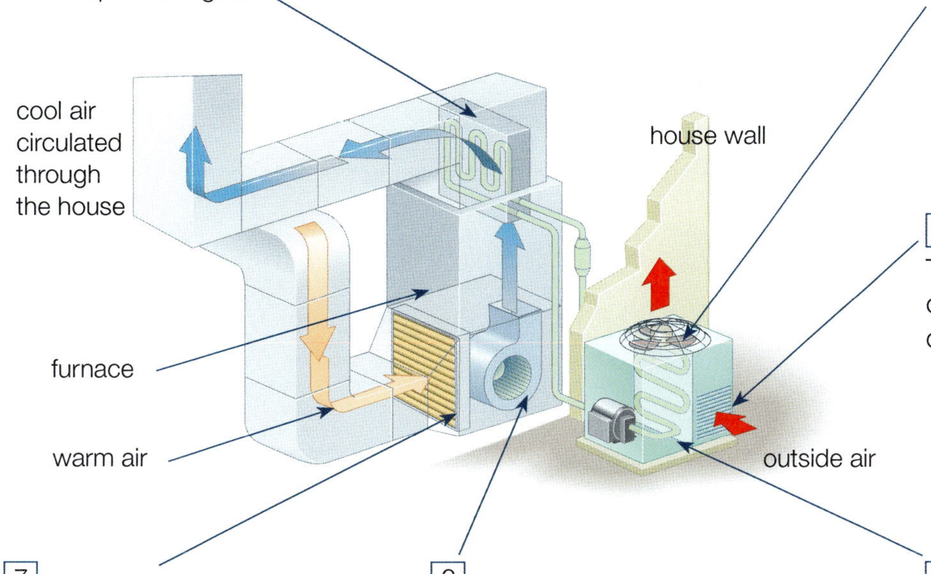

7 ______________________
This cleans the air.

6 ______________________
This moves air through the system inside the building.

5 ______________________
This changes the refrigerant into a high-temperature gas.

There are eight words, but only seven descriptions. Label the missing part. ______________________

23 3 Maintaining an air conditioning system

A supervisor is giving his apprentices tips about how to keep the air conditioning working properly. Listen to the tips and answer these questions.

1 How often should the air filter be cleaned or replaced?

____

2 What becomes very dirty and must be cleaned at least once a year?

____

3 How can they be cleaned?

____

IT'S YOUR TURN
You see this blog on how to maintain an air conditioning system and you decide to help by replying.

How do I maintain an air conditioning system?
By Ross Richardson
Trainee plumber
Question: How do I maintain an air conditioning system? I was ill the day we learned about it in class. ☹

24 Applying for a job

This unit is about:
- a letter of application
- a CV (curriculum vitae)

Competences:
- reading a letter of application
- writing a CV (curriculum vitae)

1 A letter of application

Once you have finished your apprenticeship, the next step is finding a job. You can find jobs on the Internet, in newspapers and in trade magazines. Even when you apply for a job online or by email, you should always write a short letter of application.
Read the letter below, then complete the sentences 1–6 with the missing details.

Thomas Esser • Schillerstr. 14 • 70175 Stuttgart • Tel: +49 711 71707

Ms F. Cunningham
Personnel Manager Ventifix PLC
PO Box 223987
Birmingham
Great Britain

23 February 20..

Application for the post of plumber

Dear Ms Cunningham
I am writing to apply for the post of plumber with Ventifix, which was advertised in the Coventry Evening Telegraph on 17th February. Pleased find my CV enclosed, which I hope will be of interest to you.
As you will see, I have recently finished my training as a plumber in Sindelfingen and I am now looking for full-time employment. I am applying for this post because I would like to gain experience of working in another country. I think that a post with your firm would be an interesting challenge for me at the beginning of my career.
I feel that I am well-suited to the post as I am hard-working and highly motivated. Moreover, I am sure that my extensive training and work experience will ensure that I am effective in this position.
I also speak and write good English.
I hope that I have included all the information in my CV that you might need. However, should you require further details, please do not hesitate to contact me at any time.
I look forward to hearing from you soon.
Yours sincerely
Thomas Esser

Enc.
CV, photocopies of qualifications

Tip
- Punctuation after salutation and closing must be the same.
- Always start with a capital letter after the salutation.
- Use "enclosed" and "Enc." when you send your documents by post. Use "attached" when you send your documents by email.

1 The letter is from ____________ to ____________ of Ventifix PLC.
2 Thomas found out about the job from an ____________ in the ____________.
3 Ventifix PLC is looking for ____________.
4 Thomas is looking for a job that offers ____________.
5 Thomas is applying for this job because ____________.
6 Thomas thinks he is the right person for the job because he ____________.

2 A curriculum vitae

Here are the most important things to remember when writing a CV in English.

Tip

BE curriculum vitae
AE résumé

- List details of your education, training and work experience in reverse order.
- Don't sign and date your CV.
- Your CV should never be longer than two pages.
- Include any skills and experience you have gained in addition to your formal training.
- Include details about your hobbies and interests.
- The information in your CV should be easy to find: organize it into logical blocks and give each block a heading.
- Don't include copies of qualifications or a photograph of yourself unless you are actually asked to do so.

★ a) Use the words in the box to fill the gaps in this CV.

apprenticeship • Available • Computer • Date of birth • driving • Football • Place of birth

Thomas Esser

PERSONAL DETAILS

Name	Thomas Esser
Address	Schillerstr. 14, 70175 Stuttgart
____________ 1	5 August 1997
____________ 2	Stuttgart
Telephone/Fax	+49 711 71707
Email address	t.esser@e-online.de

EDUCATION

Berufliches Schulzentrum 1, Sindelfingen

TRADE QUALIFICATIONS AND WORK EXPERIENCE

2006	Vocational Qualification in measurement, test, and control systems
2012–2016	Plumbing ____________ 3 with Schenker Meisterbetrieb, Stuttgart

OTHER SKILLS AND INTERESTS

Qualifications	General Commercial Certificate in Business Studies, vocational English
IT skills	____________ 4 course at evening school: Word, Excel, Outlook, Photoshop, and general Internet skills
Hobbies	Cycling
	____________ 5 – member of SC Kickers Amateur Football Club
	Digital photography
Other	Full clean ____________ 6 licence

REFERENCES

____________ 7 on request

★★ b) Now write your own CV.

Extra practice: Units 21–24

1 Choosing a radiator

Here is a blog entry about the different types of radiators.
Work with a partner. Read the blog entry and fill in the missing words from the words in the box.

cast iron • cheap • efficient • expensive • flat • room • sizes • work

What is the difference between the main types of radiators?
By Brian Mackenzie
Trainee plumber

Question: What is the difference between the main types of radiators?

Answer:

Radiators all ______________ 1 the same way but there are four main types. Steel radiators are ______________ 2 but they rust quite quickly. Radiators made of ______________ 3 last longer. These radiators take some time to warm up, but will hold the heat long after being turned off. Pipework radiators are more ______________ 4 but they also ______________ 5 to run and to buy. Panel radiators, that is, ______________ 6 radiators, are the most flexible because you can get them in different ______________ 7. It is important that you choose the right size according to the size of the ______________ 8 that you want to heat.

2 Group project

Get together in groups of three. Make a poster about the most important aspects of solar power.

Include the following information:

- Helping the environment
- Clean energy
- Alternative energy
- Cost

3 A problem of humidity

You have written to your landlord Martin Frazer complaining about the humidity in your apartment. The next day you get a reply.

★★ a) **Read the letter and then write your original letter of complaint.**

Dear …
I am very sorry that your apartment is showing signs of humidity. I agree with you that we could install air conditioning. You said you thought it was cheap. But actually air conditioning is extremely expensive.
I agree with you when you say that the humidity is probably caused by poor air quality. You say you can't leave the windows open all day because you go to work. But you realize that opening the windows for 10 minutes each morning will improve the air quality, don't you?
As you suggested in your letter, I think it would be a good idea if I came by and looked at the problem myself. You suggested next Saturday. If it is OK with you, I would like to come at 4 o'clock in the afternoon.
Please let me know if this suits you.
Yours sincerely
Martin Frazer

★★★ b) **Now write your reply to Mr Frazer's letter.**

4 A new job

You want to apply for a job but your colleague doesn't understand why.
Work with a partner. Use the help below to work out a dialogue. Present it to the class.

You	Your colleague
Erklären Sie Ihrer Kollegin / Ihrem Kollegen, dass Sie eine neue Stelle gefunden haben und wo Sie sie entdeckt haben (Fachzeitschrift, Internet …). →	Fragen Sie, wie die Firma heißt und wo sie ist.
← Beantworten Sie die Frage entsprechend. →	Fragen Sie, was anders ist bei der neuen Stelle.
← Sagen Sie, was anders ist. Andere Fertigkeiten wie z. B. … →	Bitten Sie um mehr Informationen.
← Sagen Sie, warum die neue Stelle Ihrer Meinung nach sehr gut zu Ihnen passt und besser als die derzeitige Stelle ist. →	Fragen Sie, ob die Kollegin / der Kollege Hilfe bei dem Lebenslauf für die Bewerbung braucht.
← Bejahen Sie die Frage und bedanken Sie sich.	

▶▶▶ ICH KANN

	Seite/Übung
☐ Ich kann eine eingeübte, kurze und einfache Präsentation zu einem Thema vortragen, das mir vertraut ist. Ich kann dabei einfache Nachfragen beantworten, wenn ich um Wiederholung bitten kann. A2	57/ It's your turn
☐ Ich kann meine Meinungen, Pläne, Absichten und Ziele darlegen und einfach begründen. B1	55/ It's your turn
☐ Ich kann ein Gespräch über mir vertraute Themen beginnen, in Gang halten und beenden. B1	56/1b
☐ Ich kann kurzen und deutlich gesprochenen Texten und Tonträgern die wesentlichen Informationen entnehmen, wenn es um vorhersehbare alltägliche oder berufsspezifisch vertraute Dinge (z. B. *types of radiators*) geht. A2	54/1
☐ Ich verstehe das Wichtigste von Radiosendungen und Tonaufnahmen, wenn sie aktuelle Ereignisse oder meine Interessen betreffen. B1	54/1
☐ Ich verstehe Arbeitsanleitungen in meiner Ausbildung und am Arbeitsplatz und schriftliche Anmerkungen zu meinen eigenen Arbeiten. A2	55/3
☐ Ich kann Informationen in verschiedenen längeren Texten oder Textteilen finden, die ich brauche, um eine bestimmte Aufgabe zu lösen. B1	58/1
☐ Ich kann einfache beschreibende Texte zu grundlegenden Sachverhalten oder Aufgabenstellungen aus meinem Fachgebiet verfassen. B1	58/1c
☐ Ich kann einen einfachen Lebenslauf und ein einfaches Bewerbungsschreiben verfassen. B1	61/2b

Wordlist

A

abroad	*im Ausland, ins Ausland*
acceptable	*annehmbar, akzeptabel*
access	*Zugang*
to activate	*aktivieren, einschalten*
adaptor	*Anschluss-, Zwischenstück*
to add	*hinzufügen*
addition	*Hinzufügung, Beimischung*
adhesive tape	*Klebeband*
to adjust	*einstellen, regeln*
adjustable pipe wrench	*verstellbare Rohrzange*
adjustable spanner	*verstellbarer Schraubenschlüssel*
adjustment	*Anpassung, Korrektur*
advanced	*fortschrittlich, modern*
advice	*Ratschlag*
air circulation	*Luftzirkulation*
air conditioning	*Klimaanlage*
airlock	*Luftblase*
air vessel	*Luftbehälter, Windkessel*
Allen key	*Inbusschlüssel, Innensechskantschlüssel*
to allow	*lassen, erlauben*
alloy	*Legierung*
alternately	*abwechselnd*
amount	*Menge*
angle grinder	*Winkelschleifer, Flex*
anticlockwise	*gegen den Uhrzeigersinn*
appearance	*Aussehen, Äußeres*
appliance	*Gerät*
application	*Bewerbung*
to apply	*auftragen; sich bewerben*
appointment	*Termin*
apprentice	*Auszubildende/r, Lehrling*
apprenticeship	*Ausbildung*
area	*Fläche, Bereich, Gebiet*
to assemble	*zusammensetzen, -bauen, montieren*
to attach	*verbinden mit, anschließen*
attached	*beigefügt*
auger	*Erdbohrer, Förderschnecke*
available	*erhältlich, verfügbar*
axial face seal	*Achsenabdichtung*

B

backing paper	*Schutz-, Trägerpapier*
back-up	*Unterstützung; unterstützend*
balance	*Gleichgewicht*
to balance	*im Gleichgewicht halten, ausgleichen*
bar of soap	*(ein Stück) Seife*
barrier	*(Schutz)barriere*
base	*Unterseite, Boden*
basin wrench	*(verstellbare) Standhahnzange*
bath mixer	*Mischbatterie (im Bad)*
bathroom	*Badezimmer*
bathroom cabinet	*Badezimmerschrank*
bathroom furnishings	*Badeinrichtung*
bathtub	*Badewanne*
bend	*Bogen*
to bend	*biegen*
bib tap	*gekrümmter Hahn mit Schlauchverbindung*
(saw) blade	*Sägeblatt*
to blast	*(mit Wucht) schießen*
to bleed	*entlüften*
bleed valve	*Entlüftungsventil*
to block	*verstopfen, blockieren*
blockage	*Verstopfung*
blocked	*blockiert, verstopft*
blower	*Gebläse*
blunt-jaw pipe wrench (pliers)	*Chrom-, Armaturenzange*
body protection	*Körperschutz*
boiler	*Heißwasserspeicher, Boiler*
bolt hole	*Schraubenbohrung, Schraubenloch*
to borrow	*ausleihen, borgen*
boss branch	*Doppel-T-Stück*
bowl	*Toilettenbecken, -schüssel*
bracket	*Klammer, Träger*
branch drain	*Sammelanschlussleitung*
branch pipe	*Stockwerksleitung*
brass	*Messing*
briskly *(adv.)*	*schnell*
bronze	*Bronze*
brush	*Bürste, Pinsel*
bubble	*Blase*
to bubble	*Blasen bilden*
bucket	*Eimer, Kübel*
building	*Gebäude*
burr	*Grat, raue Kante*
busy	*beschäftigt*
button	*Knopf*

C

cable	*Kabel*
cap	*Verschluss, Deckel*
cap nut	*Überwurf-, Kapselmutter*
capstan handle	*Drehkreuzhandgriff*
career objective	*Berufsziel*
careful(ly)	*vorsichtig*
cartridge (of tap)	*Kartusche(ndichtung der Mischbatterie)*
casing	*Gehäuse*
casting	*Guss, Gießen*
cast aluminium	*Aluminiumguss*
cast iron	*Gusseisen*
cast metal	*Gussmetall*
to cause	*verursachen*
caustic	*ätzend*
C-clamp	*Schraubzwinge*
central heating	*Zentralheizung*
centrifugal pump	*Kreiselpumpe*
cesspool	*Jauchegrube*
cheap	*billig, preisgünstig*
checklist	*Checkliste*
check valve	*Inspektions-, Prüfventil*
chemical	*chemisch; Chemikalie*
china	*Porzellan*
chrome-plated steel	*verchromter Stahl*
chromium-nickel steel	*Chromnickelstahl*
chromium-plated	*verchromt*
circulation pump	*Umwälz-, Zirkulationspumpe*
cistern	*Wassertank, Spülkasten*
clamp	*Klammer, Zwinge, Klemmschraube*
to clamp	*festhalten, einspannen*
to clean	*reinigen, sauber machen*
cleanout (plug)	*Reinigungs-, Ausputzöffnung*
to clear	*freimachen, reinigen*
clip	*Klammer, Halter*
clockwise	*im Uhrzeigersinn*
clog	*Verstopfung*
to clog	*verstopfen*
closed	*geschlossen*
cloth	*Lappen, Tuch*
cloudy	*bewölkt*
coin	*Münze*
to compare	*vergleichen*
complaint	*Reklamation, Beschwerde*
complicated	*kompliziert*
compression joint	*Quetschverbindung*
compressor	*Kompressor*
common	*üblich*
condensation	*Kondensation*
condenser	*Kondensator*
condenser coils	*Verflüssiger*
cone	*Konus, Kegel*
to connect	*verbinden*
connection	*Verbindung*
container	*Behälter, Gefäß*
continuous-flow heater	*Durchlauferhitzer*
control	*Regler*
to control	*regeln, steuern*
controls	*Steuer(ung)*
conventional	*herkömmlich*
conversion table	*Umrechnungstabelle*
to convert	*umwandeln*
copper	*Kupfer*
corkscrew	*Korkenzieher*
corrosion	*Korrosion , Rostbildung*
coupling nut	*Spann-, Überwurfmutter*
cover (of tap)	*Abdeckung (des Wasserhahns)*
to crack	*zerbrechen*
cramped	*eng, beengt*
crawler	*Crawler (Roboter)*
crutch head handle	*krückenförmiger Handgriff*
curriculum vitae	*Lebenslauf*
curved	*kurvenförmig, geschwungen*
cushion	*Kissen*
customer	*Kunde, Kundin*
customer service	*Kundendienst*
cutting disc	*Trennscheibe*

D

damage	*Schaden, Schäden*
damaged	*beschädigt*
dangerous(ly)	*gefährlich*
dead	*tot*
to deal with	*sich beschäftigen mit*
decentralized	*dezentralisiert*
to degrade	*sich abbauen, zerfallen*
degree	*Grad*
to deliver	*liefern*
depth	*Tiefe*
to destroy	*zerstören*
to detect	*aufspüren, entdecken*
detection	*Aufspüren, Entdeckung*
to develop	*entwerfen*
device	*Vorrichtung, Gerät*
to dig up	*aufgraben*
to differ	*sich unterscheiden*
difficult	*schwer, schwierig*
dirt	*Schmutz*
dirty water	*Abwasser*

disc guard	*Scheibenschutz*
to disconnect	*abklemmen, trennen, herausziehen, abmontieren*
dishwasher	*Geschirrspülmaschine*
disposal	*Beseitigung, Entsorgung*
to distribute	*verteilen*
to divide	*dividieren*
double branch	*Doppel-Abzweig*
double-handed pipe bender	*Zweihandrohrbieger*
drain	*Abfluss(rohr)*
to drain	*ablaufen lassen, leeren*
drain cleaner	*Rohrreiniger*
drain screw	*Wasserablassschraube*
drain snake	*Rohrreinigungsspirale*
drainage	*Abfluss(rohre), Entwässerung, Abwasser, Kanalisation*
drainage grate	*Abflussgitter, -sieb*
drainage pipe	*Abfluss-, Ablaufrohr, Abwasser-, Entwässerungsleitung*
drainpipe	*Regenrohr*
to draw (into)	*(ein)ziehen*
drill bit	*Bohrer, Bohreinsatz*
drinking water	*Trinkwasser*
drip	*Tropfen*
to drive	*steuern, antreiben*
dry	*trocken*
duct tape	*Klebeband*
durability	*Haltbarkeit*
dust	*Staub*
E	
earplug	*Ohr(en)stöpsel*
ear protectors	*Gehörschutz*
to ease open	*behutsam öffnen*
eaves *(pl.)* (of a roof)	*Dachvorsprung*
efficiency	*Leistungsfähigkeit, Wirksamkeit*
elbow	*Winkel*
electric drill	*Elektrobohrer*
electric pipe bender	*elektrischer Rohrbieger*
electrical tape	*Isolierband*
electric(al) outlet	*Steckdose*
electricity	*Elektrizität, Strom*
to eliminate	*beseitigen*
emergency	*Notfall, Not-*
emery paper	*Schmirgel-, Schleifpapier*
enclosed	*beigefügt, in der Anlage*
to engage in	*eingreifen in, einrasten*
environment	*Umwelt, Umgebung*
equal-size	*gleich groß*
equipment	*Geräte, Ausrüstung*
equipped with	*ausgestattet mit*
to escape	*entweichen*
essential	*unerlässlich*
estimate	*Kostenvoranschlag*
to estimate	*schätzen*
evaporator coils	*Verdampfer*
evenly *(adv.)*	*gleichmäßig*
exception	*Ausnahme*
to expand	*erweitern, vergrößern*
expensive	*teuer*
experience	*(Berufs-)Erfahrung*
exposed	*freiliegend*
exposure (to)	*Aussetzung, ausgesetzt sein*
eye protection	*Augenschutz*
F	
fabric	*Stoff*
fail-safe	*störungssicher*
false wall	*Zwischenwand*
fan	*Gebläse, Ventilator*
faucet *(AE)*	*Wasserhahn*
feature	*Merkmal, Kennzeichen*
fibre	*Faser*
field	*Feld, Gebiet, Bereich*
file	*Feile*
to fill	*füllen*
filter	*Filter*
to filter	*filtern*
fire sprinkler system	*Sprinkleranlage*
to fit	*anbringen, installieren, einbauen*
fitting	*Fitting*
to fix	*befestigen*
fixtures	*Installation(en), Armaturen*
flame	*Flamme*
flat	*Wohnung; flach*
flexible	*flexibel*
flexible drainrod	*(flexible) Spirale*
flexible tubular guide	*flexibles Führungsrohr*
floor-mounted toilet	*Standtoilette*
flow (of water)	*(Wasser-)Durchlauf*
to flush	*spülen*
flushing outlet	*Spülöffnung, -auslass*
flux	*Fluss-, Schmelzmittel*
foam tube	*Schaumstoffrohr, Schaumstoffummantelung*
foam wrap	*Schaumstoffummantelung*
footswitch	*Fußschalter*
frame	*Rahmen*
fresh	*frisch*
fridge	*Kühlschrank*
to free	*befreien*
to freeze	*einfrieren*
fuel	*Brenn-, Kraftstoff*
G	
galvanized steel	*verzinkter Stahl*
gas bottle	*Gasflasche*
gas meter	*Gaszähler*
gasket	*Dichtung(sring)*
gate valve	*Absperrventil*
gauge	*Messgerät*
to get on (well) with sb	*mit jdm. (gut) auskommen*
glazed china	*glasiertes Porzellan*
gliding paste	*Gleitpaste*
gloves	*Handschuhe*
goggles	*Schutzbrille*
gravity-fed	*durch Schwerkraft betrieben*
greasy	*fettig*
grinder	*Schleifmaschine, Winkelschleifer*
grinding disc	*Schleifscheibe*
to grip	*greifen*
(guide) groove	*(Führungs-)Rille*
guard	*Schutz*
guide	*Anleitung*
gully	*Gully*
H	
hacksaw	*Bügelsäge*
handbasin	*Handwaschbecken*
hand-held shower	*Handdusche*
handicapped	*behindert*
handle	*(Hand-)Griff*
handwheel	*Handgriff, Drehgriff*
handy	*praktisch, nützlich*
harmful	*schädlich*
headgear (of tap)	*Kopf-, Oberteil*
head for heights	*schwindelfrei*
health	*Gesundheit*
hearing	*Gehör*
hearing protection	*Gehörschutz*
heater	*Heizgerät*
heating system	*Heizanlage, Heizungsanlage*
heat	*Hitze, Wärme*
heat loss	*Wärmeverlust*
heat output	*Heizleistung*
heavy-duty hose	*strapazierfähiger Wasserschlauch*
height	*Höhe*
hexagonal	*sechseckig*
hidden	*versteckt*
hole	*Loch*
hooked	*hakenförmig*
hopper head	*Einlauf-, Ablauftrichter*
horizontal(ly)	*horizontal, waagerecht*
horizontal inlet	*horizontaler Zulauf*
hose	*Schlauch*
hose bib	*(Garten-)Schlauchverbindung*
hose clamp	*Schlauchklemme, Schelle*
hot-water storage cylinder	*Warmwasserspeicher*
house drain	*Hausentwässerung, -anbindung*
household water pump	*Hauswasserwerk*
housing	*Gehäuse*
humidity	*Luftfeuchtigkeit*
hydraulic	*hydraulisch*
hydro-jetter	*Wasserstrahler*
I	
impeller	*Schaufelrad*
immersion heater	*Tauchheizkörper*
impolite	*unhöflich*
to improve	*verbessern*
to include	*beilegen, beinhalten, umfassen*
increase	*Steigerung*
inlet pipe	*Zulauf(rohr)*
to insert	*einführen*
to inspect	*überprüfen, kontrollieren*
inspection	*Überprüfung, Untersuchung*
to install	*installieren, einbauen, -setzen, anschließen*
installation	*Installation, Auf-, Einbau*
instruction	*Anleitung*
to insulate	*isolieren*
insulated	*isoliert*
insulation	*Isolierung, Isolation*
interdependent	*ineinandergreifend*
to interrupt	*unterbrechen*
iron	*Eisen*
J	
jaw (of a tool)	*Backe*
to join	*verbinden*
joint	*Verbindung(srohr), Naht(stelle)*
jumper	*Steinbohrer, Bohrmeißel, Stauchhammer*
K	
to keep sth away from	*etwas fernhalten von*
kitchen	*Küche*
kitchen sink	*Küchenspüle*
knob	*Knopf, Drehknopf*
L	
label	*Etikett, Schild(chen)*
to label	*etikettieren, kennzeichnen, benennen*
ladder	*Leiter*
to lag	*isolieren*

lagging	*Isolierung*
late	*spät*
laundry basket / bin	*Wäschekorb*
layer	*Schicht*
leaflet	*Merkblatt, Broschüre*
leak	*undichte Stelle, Leck*
to leak	*undicht sein, lecken, auslaufen*
length	*Länge*
letter of application	*Bewerbungsschreiben*
level	*waagerecht; Grad, Menge*
lever	*Hebel*
to lever out	*aushebeln*
to lift away	*entfernen, hochheben*
limit	*Grenze*
to limit	*begrenzen, beschränken*
to line	*auskleiden*
to line up with	*bündig abschließen mit*
lining	*Auskleidung*
locking nut	*Gegenmutter*
lockshield valve	*Rücklaufverschraubung*
long-lasting	*langlebig*
loose	*lose*
to loosen	*lockern, lösen*
low	*niedrig*
low-gauge	*niedriger Pegel*
to lubricate	*schmieren*

M

main	*Haupt-*
main drain	*Hauptrohr, -kanal (für Abwasser)*
main vent stack	*Hauptabluftkamin*
mains, water mains	*Wasserhauptleitung*
to maintain	*warten; instandhalten*
maintenance	*Wartung, Pflege*
manhole cover	*Gullydeckel*
manometer	*Manometer*
to mark	*markieren*
mass	*Masse, Gewicht*
master plumber	*Installateurmeister/in*
materials	*Materialien*
maths	*Mathe, Mathematik*
to measure	*messen*
to melt	*schmelzen*
mess	*Durcheinander, Unordnung*
message, to leave a message on voicemail	*auf die Voice-Mailbox sprechen*
metallic	*metallisch, Metall-*
metal spiral	*Metallspirale*
mixer tap	*Mischbatterie, Mischarmatur*
mixer tap for sinks	*Spülenmischarmatur, Mischbatterie für Spülen*
mobile	*Handy*
moisture	*Feuchtigkeit*
molten iron	*geschmolzenes Eisen*
motorized	*motorisiert*
motor-pump shaft	*Motorpumpenwelle*
mould	*Gussform*
to mount	*installieren*
to move	*bewegen*
moving parts	*bewegliche Teile*
multi-point system	*System mit Mehrfachanschluss*
to multiply	*multiplizieren*
multistage	*Multi-, mehrstufig*

N

necessary	*nötig*
noise	*Geräusch*
noisy	*laut*
nozzle	*Düse, Strahltülle*
nut	*Mutter, Schraube*

O

O-ring	*Dichtungsring, Rundring*
obstruction	*Hindernis, Verstopfung*
occasional(ly)	*gelegentlich*
to occur	*vorkommen*
offset	*Absatzstück*
oily	*ölig, fettig*
to operate	*bedienen, betreiben*
operating pressure	*Betriebsdruck*
to order (a part)	*ein (Ersatz)teil bestellen*
out of service	*außer Betrieb*
outflow pipe	*Ablaufrohr*
outlet pipe	*Auslauf-, Abflussrohr*
overalls	*Overall, Arbeitsanzug*
overflow	*Überlauf*
to overflow	*überlaufen*
to overheat	*überhitzen*
overlap	*Überlappung*
to overlap	*überlappen (lassen)*
oxidation	*Oxidation, Oxidierung*

P

packing tape	*Dichtungsband*
pad	*Polster*
to pad	*polstern*
panel radiator	*Flach-, Plattenheizkörper*
part-time	*Teilzeit*
paste	*Paste*
to peel off	*abziehen*
pillar tap	*einfacher Wasserhahn mit Handgriff oben*
pipe-bending machine	*Rohrbiegegerät*
pipe cutter	*Rohr(ab)schneider*
pipefitter	*Rohrleger/in*
pipe fitting	*Rohrfitting*
pipe fittings	*Rohrverbindungen*
pipe inspection	*Rohrüberprüfung, -untersuchung*
pipe vice	*Rohrschraubstock*
pipe wrench	*(Einhand-)Rohrzange, Rohrschlüssel, Wasserpumpenzange*
pipe-threading machine	*Gewindeschneidmaschine*
pipework	*Rohrleitung(en)*
pipework radiator	*Bad-, Designheizkörper*
piping (systems)	*Rohrleitung(en)*
pliers *(pl)*	*Zange*
plug	*Stöpsel*
to plug	*(zu)stopfen, (ver)schließen*
to plug into	*einstöpseln, anschließen an*
plughole	*Abfluss, Abflussloch*
plumber	*(Gas-, Wasser-)Installateur/in, Klempner/in*
plumbing company	*Sanitärbetrieb*
plumbing kit	*Werkzeugkasten*
plumbing plan	*Rohrleitungsplan*
plumbing supplies	*Sanitärbedarf*
plumbing system	*Rohrleitungen, Rohrleitungssystem*
plunger	*Saugglocke*
poisonous	*giftig*
to polish	*polieren*
polite	*höflich*
politeness	*Höflichkeit*
to pollute	*(die Umwelt) verschmutzen*
poor (air quality)	*schlecht(e) (Luftqualität)*
porcelain	*Porzellan*
portable	*tragbar*
portable hydraulic pipe bender	*tragbarer hydraulischer Rohrbieger*
to pour	*gießen*
powder	*Staub*
power	*Kraft, Energie, Strom*
power cable	*Stromkabel*
power tool	*elektrisch oder pneumatisch angetriebenes Werkzeug*
precaution	*Vorsichtsmaßnahme*
to pre-set	*voreinstellen*
pressure	*Druck*
pressure casing	*Druckgehäuse, -verkleidung*
pressure gauge	*Druckmesser, Manometer*
pressure gland	*Druckstutzen*
pressure regulator	*Druckregler*
pressure sensor	*Drucksensor*
pressure switch	*Druckschalter*
pressure vessel	*Druckbehälter, -kessel*
to prevent	*verhindern*
process	*Verfahren(sweise), Prozess*
to produce	*erzeugen*
propane torch	*Propanschweißbrenner, -lötlampe*
properly	*richtig, ordnungsgemäß*
to protect	*schützen*
protective (headgear)	*Schutz-, (Kopfschutz)*
to provide	*liefern*
public	*öffentlich*
pump	*Pumpe*
to pump	*pumpen*
pump casing	*Pumpengehäuse*
pump housing	*Pumpengehäuse*
punctual	*pünktlich*
putty	*(Dicht-)Kitt*

Q

questionnaire	*Fragebogen*
quiet	*ruhig, leise*
to quote (a price)	*(Preis) angeben, anbieten*

R

radiator	*Heizkörper*
radiator spanner	*Heizkörperschraubenschlüssel*
rag	*(Scheuer-, Putz-)Lappen*
range	*Bandbreite, Auswahl*
razor blade	*Rasierklinge*
to reach	*erreichen*
reamer	*Entgrater*
to reconnect	*wieder verbinden*
to reduce	*reduzieren*
to refill	*neu befüllen*
to refit	*wieder anbringen*
refrigerant	*Kältemittel*
to refuse	*ablehnen*
to regulate	*regeln*
regulation	*Regelung*
to release	*entlassen, freisetzen*
to reline	*wieder auskleiden*
to remind sb	*jdn erinnern*
to remove	*entfernen, beseitigen*
to renovate	*renovieren*
repair	*Reparatur*
to repair	*reparieren*
to replace	*ersetzen, austauschen, wieder einsetzen*
replacement part	*Ersatzteil*
request	*Bitte*
resin	*Harz*
respirator	*Atemschutz(maske), Mundschutz*
responsible	*verantwortlich*
résumé *(AE)*	*(tabellarischer) Lebenslauf*
retaining nut	*Überwurfmutter*
retaining screw	*Halteschraube*
to return sth	*etwas zurückbringen*
reverse order	*umgekehrte Reihenfolge*
rigid	*fest*

riser	*Steigleitung*
robust	*robust*
rocking	*Wackeln*
rod	*Stange, Stab*
to rotate	*rotieren (lassen)*
rubber	*Gummi*
rude	*unhöflich*
to run dry	*(bei einer Pumpe) ohne Wasser laufen, trockenlaufen*
rust	*Rost*
to rust	*rosten*
rust-free	*rostfrei*

S

safety	*Sicherheit*
safety boots	*Sicherheitsschuhe*
safety helmet	*Schutzhelm*
safety procedures	*Sicherheitsvorschriften*
safety rule	*Sicherheitsregel*
safety sign	*Sicherheitsschild*
salmon	*lachsfarben*
to save money	*Geld sparen*
scaffolding	*(Bau-)Gerüst*
scales	*Waage*
scaling	*Verkalkung, Kesselstein-bildung*
scratch	*Kratzer*
to scratch	*(zer)kratzen*
screw	*Schraube*
to screw (onto)	*(an)schrauben*
screwdriver	*Schraubenzieher, -dreher*
seal	*Dichtung*
to seal	*(ab)dichten, verschlie-ßen, versiegeln*
sealing	*Dichtung(s-)*
section	*Teil*
to secure	*befestigen*
self-adhesive	*selbstklebend*
sensor	*Sensor, (Mess-)Fühler*
separate	*getrennt*
septic tank	*Kleinkläranlage*
serious(ly)	*ernsthaft*
to service	*warten (Maschine)*
service valve	*Wartungsventil*
to set	*(ein)stellen*
setscrew	*Stell-, Feststell-, Befesti-gungsschraube*
severe frost	*strenger Frost*
sewer gas	*Abwassergas*
sewer (pipe)	*Abwasser(leitung)*
shade	*Schatten*
to shape	*formen*
shaft hole	*Achsloch*
to share	*teilen*
shim	*Keil, Unterlegstück, Ausgleichscheibe*
shoe (fitting)	*Auslaufstück*
shower	*Dusche*
shower head	*Duschkopf*
shut-off valve	*Absperrventil*
side handle	*Seitengriff*
single branch	*Einfach-Abzweig*
single-control tap	*Einhebelwasserhahn*
single-point	*Einzelanschluss*
sink	*Spüle, Ausguss, Spülbecken*
sink mixer	*Mischbatterie (über der Spüle)*
site (building site)	*Baustelle*
skill	*Fertigkeit, Fähigkeit*
size	*Größe*
sleeve	*Manschette*
to slide up	*hochrutschen*
slim	*dünn, schmal*
to slip down	*heruntergleiten lassen*
to slip in/under	*einführen*
to slow down	*verlangsamen*
to smell	*riechen*
snake	*Schlange*
to soak	*einweichen, durch-tränken*
soil and waste stack	*Fallleitung*
soil stack	*Fallleitung*
solar collector	*Sonnenkollektor*
solar panel	*Sonnenkollektor*
solar power	*Sonnenkraft, Solarstrom*
solder	*Lötzinn*
to solder	*löten*
soldering wire	*Lötdraht*
solid	*fest, hart*
to solve	*lösen*
source	*Quelle*
spanner	*Schraubenschlüssel*
spiral	*Spirale*
spirit level	*Wasserwaage*
to split	*zerpringen*
sponge	*Schwamm*
spout	*Auslauf*
spring	*Quelle*
square	*viereckig, quadratisch*
stage chamber	*Stufenkammer*
stainless steel	*rostfreier Stahl*
steel	*Stahl*
steel wool	*Stahlwolle*
stiff	*fest, schwergängig*
stopcock	*Haupt-, Absperrhahn*
storage battery	*Speicherbatterie*
storage heater	*Warmwasserspeicher*
storage tank	*Wasservorratsbehälter*
to store	*speichern*
to subdivide	*unterteilen*
to subtract	*subtrahieren*
successful	*erfolgreich*
suction	*Ansaug-*
superior	*höherwertig, überlegen*
supervisor	*Vorgesetzte/r*
(plumbing) supplies	*(Installations-)Bedarf*
supply pipe (line)	*Versorgungsleitung*
to surround	*umgeben, ummanteln*
switch	*Schalter, Knopf*
to switch on/off	*ein-/ausschalten*

T

tap	*(Wasser-)Hahn*
to tap	*(an)tippen, klopfen*
tap washer	*Beilag-, Unterleg-scheibe, -ring, Dichtung(sscheibe)*
technician	*Techniker/in*
temperature regulator	*Temperaturregler*
thermoplastic	*thermoplastisch*
thermostat mixer (tap)	*Mischbatterie mit Thermostat*
thickness	*Dicke, Stärke*
thread	*Gewinde*
to thread	*Gewinde schneiden*
threaded pipe	*Rohr mit Gewinde*
tick	*Häkchen*
to tick	*abhaken*
to tighten	*anziehen, festschrauben*
tile	*Fliese*
tin	*Zinn, Blech*
tool	*Werkzeug*
to touch	*berühren (lassen)*
tough	*hartnäckig*
to toughen	*härter machen, verstärken*
towel	*Handtuch*
towel rail	*Handtuchhalter*
toxic material	*Giftstoff*
tradesman	*Handwerker*
trap	*Siphon(winkel), Geruchsverschluss*
trapped (air)	*eingeschlossen(e) (Luft)*
tubing	*Röhren(system)*
tubular	*röhrenförmig*
to turn on/off	*ein-/ausschalten*
to twist	*drehen*
two-grip tap	*Hahn mit zwei Ventilen*

U

U-bend trap	*Doppelbogensiphon*
to unblock	*Verstopfung beseitigen, von einer Verstopfung befreien*
to unclog	*Verstopfung beseitigen, von einer Verstopfung befreien*
underneath	*unterhalb*
to undo	*entfernen, abschrauben*
union nut	*(Rohr-)Verschraubung*
unpleasant smells	*unangenehme Gerüche*
to unscrew	*abschrauben*
upright	*senkrecht*
upside down	*auf den Kopf gestellt, umgedreht*
to use	*benutzen, verwenden*
used water	*Abwasser*

V

valve	*Ventil*
van	*Lieferwagen, Transporter*
vapour lock	*Dampfblase*
to vary	*variieren, unterschiedlich sein*
vent	*Be-, Entlüftung*
to vent	*entlüften*
ventilation	*Lüftung, Belüftung*
vent screw	*Be-, Entlüftungs-schraube*
vertical(ly)	*vertikal, senkrecht*
vessel	*Behälter, Kessel*
volume	*Volumen, Inhalt, Menge, Umfang*

W

washbasin	*Waschbecken*
washer	*Beilag-, Unterleg-scheibe, -ring, Dichtung(sscheibe)*
washer-retaining nut	*Halteschraube der Dichtung*
washing machine	*Waschmaschine*
waste	*Abfall, Abwasser, Schwemmstoffe*
waste pipe	*Abflussrohr*
water flow	*Wasser(durch)fluss*
water heater	*Wasserheizer*
water mains	*Wassernetz*
water meter	*Wasseruhr*
water pressure	*Wasserdruck*
water pump pliers	*Wasserpumpenzange*
water-resistant	*wasserabweisend*
water supply	*Wasseranschluss, -leitung, -versorgung*
water tray	*Wasser(auffang)becken*
to wear	*tragen (Kleider)*
to weld	*schweißen*
well	*Brunnen*
wet	*nass*
wheel	*Rad*
wing nut	*Flügelmutter*
to wipe (clean)	*abwischen*
wire	*Draht*
wire brush	*Drahtbürste*
wire wool	*Drahtwolle*
wiring	*Verdrahtung, Verkabe-lung, Kabel*
wooden	*hölzern*
worn out	*abgenutzt, verschlissen*

JOB MATTERS
SANITÄR- UND KLIMATECHNIK

Verfasser:	Wolfram Lepka, Peter Oldham, Ingrid Preedy, Ken Thomson
Projektleitung:	Kerstin Adlesgruber, Simone Conrad
Verlagsredaktion:	Verena Reiter, Kari-ann Warnakulasuriya
Redaktionelle Mitarbeit:	Christine House
Umschlaggestaltung:	vitaledesign, Berlin
Layout:	Julia Dresch, Alois Kandler
Technische Umsetzung:	Breiner Satzstudio; Oxana Rödel, Absatz DTP-Service, Teltow
Coverfoto:	Shutterstock / nikkytok
Illustrationen:	Oxford Designers & Illustrators

Erhältlich sind auch:

Handreichungen für den Unterricht als Download	ISBN 978-3-06-451334-1
Audios als Download	ISBN 978-3-06-451335-8

Soweit in diesem Lehrwerk Personen fotografisch abgebildet sind und ihnen von der Redaktion fiktive Namen, Berufe, Dialoge und Ähnliches zugeordnet oder diese Personen in bestimmte Kontexte gesetzt werden, dienen diese Zuordnungen und Darstellungen ausschließlich der Veranschaulichung und dem besseren Verständnis des Inhalts.

www.cornelsen.de

Die Webseiten Dritter, deren Internetadressen in diesem Lehrwerk angegeben sind, wurden vor Drucklegung sorgfältig geprüft. Der Verlag übernimmt keine Gewähr für die Aktualität und den Inhalt dieser Seiten oder solcher, die mit ihnen verlinkt sind.

1. Auflage, 7. Druck 2026

Alle Drucke dieser Auflage sind inhaltlich unverändert und können im Unterricht nebeneinander verwendet werden.

Druck: Cornelsen Verlagskontor, Bielefeld

ISBN 978-3-06-451333-4

PEFC
PEFC/04-31-3871

PEFC-zertifiziert
Dieses Produkt stammt aus nachhaltig bewirtschafteten Wäldern und kontrollierten Quellen
www.pefc.de